The Ascent of Man

HENRY DRUMMOND

COSIMOCLASSICS

NEW YORK

The Ascent of Man
Cover © 2007 Cosimo, Inc.

For information, address:

Cosimo, P.O. Box 416
Old Chelsea Station
New York, NY 10113-0416

or visit our website at:
www.cosimobooks.com

The Ascent of Man was originally published in 1894.

Cover design by www.kerndesign.net

ISBN: 978-1-60206-194-1

To give an account of Evolution, it need scarcely be remarked, is not to account for it. No living thinker has yet found it possible to account for Evolution.

—from The Ascent of Man

PREFACE.

"THE more I think of it," say s Ruskin, "I find this conclusion more impressed upon me—that the greatest thing a human soul ever does in this world is to *see* something, and tell what it *saw* in a plain way." In these pages an attempt is made to "tell in a plain way" a few of the things which Science is now seeing with regard to the Ascent of Man. Whether these seeings are there at all is another matter. But, even if visions, every thinking mind, through whatever medium, should look at them. What Science has to say about *himself* is of transcendent interest to Man, and the practical bearings of this theme are coming to be more vital than any on the field of knowledge. The thread which binds the facts is, it is true, but a hypothesis. As the theory, nevertheless, with which at present all scientific work is being done, it is as-sumed in every page that follows.

Though its stand-point is Evolution and its subject Man, this book is far from being designed to prove that Man has relations, compromising or otherwise, with lower animals. Its theme is Ascent, not Descent, It is a Story, not an Argument. And Evolution, in the narrow sense in which it is often used when applied to Man, plays little part in the drama outlined here. So far as the general scheme of Evolution is introduced—and in the Introduction and elsewhere this is done at length—the object is the important one of pointing out how its nature has been misconceived, indeed how its greatest factor has been overlooked in almost all contemporary scientific thinking. Evolution was given to the modern world out of focus, was first seen by it out of focus, and has remained out of focus to the present hour. Its general basis has never been re-examined since the time of Mr. Darwin; and not only such speculative sciences as Teleology, but working sciences like Sociology, have been led astray by a fundamental omission. An Evolution Theory drawn to scale, and with the lights and shadows properly adjusted—adjusted to the whole truth and reality of Nature and of Man—is needed at present as a standard for modern thought; and though a reconstruction of such magnitude is not here presumed, a primary object of these pages is to supply at least the accents for such a scheme.

Beyond an attempted re-adjustment of the accents there is nothing here for the specialist—except, it may be, the reflection of his own work. Nor, apart from Teleology, is there anything for the theologian. The limitations of a lecture-audience made the treatment of such themes as might appeal to him impossible; while owing to the brevity of the course, the Ascent had to be stopped at a point where all the higher interest begins. All that the present volume covers is the Ascent of Man, the Individual, during the earlier stages of his evolution. It is a study in embryos, in rudiments, in installations; the scene is the primeval forest; the date, the world's dawn. Tracing his rise as far as Family Life, this history does not even follow him into the Tribe; and as it is only then that social and moral life begins in earnest, no formal discussion of these high themes occurs. All the higher forces and phenomena with which the sciences of Psychology, Ethics, and Theology usually deal come on the world's stage at a later date, and no one need be surprised if the semi-savage with whom we leave off is found wanting in so many of the higher potentialities of a human being.

The Ascent of Mankind, as distinguished from the Ascent of the Individual, was orginally summarized in one or two closing lectures, but this stupendous subject would require a volume for itself, and these fragments have been omitted for the present. Doubtless it may disappoint some that at the close of all the bewildering vicissitudes outlined here, Man should appear, after all, so poor a creature. But the great lines of his youth are the lines of his maturity, and it is only by studying these, in themselves and in what they connote, that the nature of Evolution and the quality of Human Progress can be perceived.

HENRY DRUMMOND.

CONTENTS.

CHAPTER III.

CHAPTER IV.

CHAPTER V.

CHAPTER VI.

CHAPTER VII.

CHAPTER VIII.

CHAPTER IX.

CHAPTER X.

INTRODUCTION.

I.

EVOLUTION IN GENERAL.

THE last romance of Science, the most daring it has ever tried to pen, is the Story of the Ascent of Man. Withheld from all the wistful eyes that have gone before, whose reverent ignorance forbade their wisest minds to ask to see it, this final volume of Natural History has begun to open with our century's close. In the monographs of His and Minot, the Embryology of Man has already received a just expression; Darwin and Haeckel have traced the origin of the Animal-Body; the researches of Romanes mark a beginning with the Evolution of Mind; Herbert Spencer has elaborated theories of the development of Morals; Edward Caird of the Evolution of Religion. Supplementing the contributions of these authorities, verifying, criticising, combating, rebutting, there works a multitude of others who have devoted their lives to the same rich problems, and already every chapter of the bewildering story has found its editors.

Yet, singular though the omission may seem, no connected outline of this great drama has yet been given us. These researches, preliminary reconnais-sances though they be, are surely worthy of being looked upon as a whole. No one can say that this multitude of observers is not in earnest, nor their work honest, nor their methods competent to the last powers of science. Whatever the uncertainty of the field, it is due to these pioneer minds to treat their labor with respect. What they see in the unexplored land in which they travel belongs to the world. By just such methods, and by just such men, the map of the world of thought is filled in—here from the trac-ing up of some great river, there from a bearing taken roughly in a darkened sky, yonder from a sudden glint of the sun on a far-off mountain-peak, or by a swift induction of an adventurous mind from a momentary glimpse of a natural law. So knowledge grows; and in a century which has added to the sum of human learning more than all the centuries that are past, it is not to be conceived that some further revelation should not await us on the highest themes of all.

The day is forever past when science need apolo-gize for treating Man as an object of natural research. Hamlet's "being of large discourse, looking before and after" is withal a part of Nature, and can neither be made larger nor smaller, anticipate less nor prophesy less, because we investigate, and perhaps discover, the secret of his past. And should that past be proved to be related in undreamed-of ways to that of all other things in Nature, "all other things" have that to gain by the alliance which philosophy and theology for centuries have striven to win for them. Every step in the proof of the oneness in a universal evolutionary process of this divine humanity of ours is a step in the proof of the divinity of all lower things. And what is of infinitely greater moment, each footprint discovered in the Ascent of Man is a guide to the step to be taken next. To discover the rationale of social progress is the ambition of this age. There is an extraordinary human interest abroad about this present world itself, a yearning desire, not from curious but for practical reasons, to find some light upon the course; and as the goal comes nearer the eagerness passes into suspense to know the shortest and the quickest road to reach it. Hence the Ascent of Man is not only the noblest problem which science can ever study, but the practical bearings of this theme are great beyond any other on the roll of knowledge.

Now that the first rash rush of the evolutionary invasion is past, and the sins of its youth atoned for by sober concession, Evolution is seen to be neither more nor less than the story of creation as told by those who know it best. "Evolution," says Mr. Huxley, "or development is at present employed in biology as a general name for the history of the steps by which any living being has acquired the morphological and the physiological characters which distinguish it." [1] Though applied specifically to plants and animals this definition expresses the chief sense in which Evolution is to be used scientifically at present. We shall use the word, no doubt, in others of its many senses; but after all the blood spilt, Evolution is simply "history," a "history of steps," a "general name," for the history of the steps by which the world has come to be what it is. According to this general definition, the story of Evolution is nar-

[1] *Encyclopædia Britannica,* 9th Ed.

rative. It may be wrongly told; it may be colored, exaggerated, over or understated like the record of any other set of facts; it may be told with a theological bias or with an anti-theological bias; theories of the process may be added by this thinker or by that; but these are not of the substance of the story. Whether history is told by a Gibbon or a Green the facts remain, and whether Evolution be told by a Haeckel or a Wallace we accept the narrative so far as it is a rendering of Nature, and no more. It is true, before this story can be fully told, centuries still must pass. At present there is not a chapter of the record that is wholly finished. The manuscript is already worn with erasures, the writing is often blurred, the very language is uncouth and strange. Yet even now the outline of a continuous story is beginning to appear—a story whose chief credential lies in the fact that no imagination of man could have designed a spectacle so wonderful, or worked out a plot at once so intricate and so transcendently simple.

This story will be outlined here partly for the story and partly for a purpose. A historian dare not have a prejudice, but he cannot escape a purpose—the purpose, conscious or unconscious, of unfolding the purpose which lies behind the facts which he narrates. The interest of a drama—the authorship of the play apart—is in the players, their character, their motives, and the tendency of their action. It is impossible to treat these players as automata. Even if automata, those in the audience are not. Hence, where interpretation seems lawful, or comment warranted by the facts, neither will be withheld.

To give an account of Evolution, it need scarcely be

remarked, is not to account for it. No living thinker
has yet found it possible to account for Evolution.
Mr. Herbert Spencer's famous definition of Evolution
as "a change from an indefinite incoherent homogene-
ity to a definite coherent heterogeneity through continuous differentiations and integrations "[1]—the formula
of which the *Contemporary Reviewer* remarked that
"the universe may well have heaved a sigh of relief
when, through the cerebration of an eminent thinker,
it had been delivered of this account of itself "—is
simply a summary of results, and throws no light,
though it is often supposed to do so, upon ultimate
causes. While it is true, as Mr. Wallace affirms in his
latest work, that "Descent with modification is now
universally accepted as the order of nature in the
organic world," there is everywhere at this moment
the most disturbing uncertainty as to how the Ascent
even of species has been brought about. The attacks
on the Darwinian theory from the outside were never
so keen as are the controversies now raging in scientific circles, over the fundamental principles of Darwinism itself. On at least two main points—sexual
selection and the origin of the higher mental characteristics of man—Mr. Alfred Russel Wallace, co-discoverer with Darwin of the principle of Natural Selection though he be, directly opposes his colleague.
The powerful attack of Weismann on the Darwinian
assumption of the inheritability of acquired characters
has opened one of the liveliest controversies of recent
years, and the whole field of science is hot with controversies and discussions. In his "Germ-Plasm," the
German naturalist believes himself to have finally

[1] *Data of Ethics,* p. 65.

disposed of both Darwin's "gemmules" and Herbert Spencer's "primordial units," while Eimer breaks a lance with Weismann in defence of Darwin, and Herbert Spencer replies for himself, assuring us that "either there has been inheritance of acquired characters or there has been no evolution."

It is the greatest compliment to Darwinism that it should have survived to deserve this era of criticism. Meantime all prudent men can do no other than hold their judgment in suspense both as to that specific theory of one department of Evolution which is called Darwinism, and as to the factors and causes of Evolution itself. No one asks more of Evolution at present than permission to use it as a working theory. Undoubtedly there are cases now before Science where it is more than theory—the demonstration from Yale, for instance, of the Evolution of the Horse; and from Steinheim of the transmutation of Planorbis. In these cases the missing links have come in one after another, and in series so perfect, that the evidence for their evolution is irresistible. "On the evidence of Palæontology," says Mr. Huxley in the *Encyclopædia Britannica*, "the evolution of many existing forms of animal life from their predecessors is no longer an hypothesis but an historical fact." And even as to Man, most naturalists agree with Mr. Wallace who "fully accepts Mr. Darwin's conclusion as to the essential identity of Man's bodily structure with that of the higher mammalia and his descent from some ancestral form common to man and the anthropoid apes," for "the evidence of such descent appears overwhelming and conclusive." [1] But as to the development of the

[1] *Darwinism*, p. 451.

whole Man it is sufficient for the present to rank it as a theory, no matter how impressive the conviction be that it is more. Without some hypothesis no work can ever be done, and, as every one knows, many of the greatest contributions to human knowledge have been made by the use of theories either seriously imperfect or demonstrably false. This is the age of the evolution of Evolution. All thoughts that the Evolutionist works with, all theories and generalizations, have been themselves evolved and are now being evolved. Even were his theory perfected its first lesson would be that it was itself but a phase of the Evolution of further opinion, no more fixed than a species, no more final than the theory which it displaced. Of all men the Evolutionist, by the very nature of his calling, the mere tools of his craft, his understanding of his hourly shifting place in this always moving and ever more mysterious world, must be humble, tolerant, and undogmatic.

These, nevertheless, are cold words with which to speak of a Vision—for Evolution is after all a Vision—which is revolutionizing the world of Nature and of thought, and, within living memory, has opened up avenues into the past and vistas into the future such as science has never witnessed before. While many of the details of the theory of Evolution are in the crucible of criticism, and while the field of modern science changes with such rapidity that in almost every department the text-books of ten years ago are obsolete to-day, it is fair to add that no one of these changes, nor all of them together, have touched the general theory itself except to establish its strength, its value, and its universality. Even more remarkable than the rapidity of its conquest is the authority with which the doctrine of development has seemed to speak to the most authoritative minds of our time. Of those who are in the front rank, of those who by their knowledge have, by common consent, the right to speak, there are scarcely any who do not in some form employ it in working and in thinking. Authority may mean little; the world has often been mistaken; but when minds so different as those of Charles Darwin and of T. H. Green, of Herbert Spencer and of Robert Browning, build half the labors of their life on this one law, it is impossible, and especially in the absence of any other even competing principle at the present hour, to treat it as a baseless dream. Only the peculiar nature of this great generalization can account for the extraordinary enthusiasm of this acceptance. Evolution has done for Time what Astronomy has done for Space. As sublime to the reason as the Science of the Stars, as overpowering to the imagination, it has thrown the universe into a fresh perspective, and given the human mind a new dimension. Evolution involves not so much a change of opinion as a change in man's whole view of the world and of life. It is not the statement of a mathematical proposition which men are called upon to declare true or false. It is a method of looking upon Nature. Science for centuries devoted itself to the cataloguing of facts and the discovery of laws. Each worker toiled in his own little place—the geologist in his quarry, the botanist in his garden, the biologist in his laboratory, the astronomer in his observatory, the historian in his library, the archæologist in his museum. Suddenly these workers looked up; they spoke to one another;

they had each discovered a law; they whispered its name. It was Evolution. Henceforth their work was one, science was one, the world was one, and mind, which discovered the oneness, was one.

Such being the scope of the theory, it is essential that for its interpretation this universal character be recognized, and no phenomenon in nature or in human nature be left out of the final reckoning. It is equally clear that in making that interpretation we must begin with the final product, Man. If Evolution can be proved to include Man, the whole course of. Evolution and the whole scheme of Nature from that moment assume a new significance. The beginning must then be interpreted from the end, not the end from the beginning. An engineering workshop is unintelligible until we reach the room where the completed engine stands. Everything culminates in that final product, is contained in it, is explained by it. The Evolution of Man is also the complement and corrective of all other forms of Evolution. From this height only is there a full view, a true perspective, a consistent world. The whole mistake of naturalism has been to interpret Nature from the stand-point of the atom—to study the machinery which drives this great moving world simply as machinery, forgetting that the ship has any passengers, or the passengers any captain, or the captain any course. It is as great a mistake, on the other hand, for the theologian to separate off the ship from the passengers as for the naturalist to separate off the passengers from the ship. It is he who cannot include Man among the links of Evolution who has greatly to fear the theory of development. In his jealousy for that religion which seems to him higher than science, he removes at once the rational basis from religion and the legitimate crown from science, forgetting that in so doing he offers to the world an unnatural religion and an inhuman science. The cure for all the small mental disorders which spring up around restricted applications of Evolution is to extend it fearlessly in all directions as far as the mind can carry it and the facts allow, till each man, working at his subordinate part, is compelled to own, and adjust himself to, the whole.

If the theological mind be called upon to make this expansion, the scientific man must be asked to enlarge his view in another direction. If he insists upon including Man in his scheme of Evolution, he must see to it that he include the whole Man. For him at least no form of Evolution is scientific, or is to be considered, which does not include the whole Man, and all that is in Man, and all the work and thought and life and aspiration of Man. The great moral facts, the moral forces so far as they are proved to exist, the moral consciousness so far as it is real, must come within its scope. Human History must be as much a part of it as Natural History. The social and religious forces must no more be left outside than the forces of gravitation or of life. The reason why the naturalist does not usually include these among the factors in Evolution is not oversight, but undersight. Sometimes, no doubt, he may take at their word those who assure him that Evolution has nothing to do with those higher things, but the main reason is simply that his work does not lie on the levels where those forces come into play. The specialist is not to be blamed for this ; limitation is his strength. But when the specialist proceeds to reconstruct the universe from his little corner of it, and especially from his level of it, he not only injures science and philosophy, but may fatally mislead his neighbors. The man who is busy with the stars will never come across Natural Selection, yet surely must he allow for Natural Selection in his construction of the world as a whole. He who works among star-fish will encounter little of Mental Evolution, yet will he not deny that it exists. The stars have voices, but there are other voices; the star-fishes have activities, but there are other activities. Man, body, soul, spirit, are not only to be considered, but are first to be considered in any theory of the world. You cannot describe the life of kings, or arrange their kingdoms, from the cellar beneath the palace. "Art," as Browning reminds us,

> "Must fumble for the whole, once fixing on a part,
> However poor, surpass the fragment, and aspire
> To reconstruct thereby the ultimate entire."

II.

THE MISSING FACTOR IN CURRENT THEORIES.

But it is not so much in ignoring Man that evolutionary philosophy has gone astray; for of that error it has seriously begun to repent. What we have now to charge against it, what is a main object of these pages to point out, is that it has misread Nature herself. In "fixing on a part" whereby to "reconstruct the ultimate," it has fixed upon a part which is not the most vital part, and the reconstructions, therefore, have come to be wholly out of focus. Fix upon the wrong " part," and the instability of the fabric built upon it is a foregone conclusion. Now, although reconstructions of the cosmos in the light of Evolution are the chief feature of the science of our time, in almost no case does even a hint of the true scientific stand-point appear to be perceived. And although it anticipates much that we should prefer to leave untouched until it appears in its natural setting, the gravity of the issues makes it essential to summarize the whole situation now.

The root of the error lies, indirectly rather than directly, with Mr. Darwin. In 1859, through the publication of the *Origin of Species*, he offered to the world what purported to be the final clue to the course of living Nature. That clue was the principle of the Struggle for Life. After the years of storm and stress which follow the intrusion into the world of all great thoughts, this principle was universally accepted as the key to all the sciences which deal with life. So ceaseless was Mr. Darwin's emphasis upon this factor, and so masterful his influence, that, after the first sharp conflict, even the controversy died down. With scarce a challenge the Struggle for Life became accepted by the scientific world as the governing factor in development, and the drama of Evolution was made to hinge entirely upon its action. It became the " part" from which science henceforth went on " to reconstruct the whole," and biology, sociology, and teleology, were built anew on this foundation.

That the Struggle for Life has been a prominent

actor in the drama is certain. Further research has only deepened the impression of the magnitude and universality of this great and far-reaching law. But that it is the sole or even the main agent in the process of Evolution must be denied. Creation is a drama, and no drama was ever put upon the stage with only one actor. The Struggle for Life is the "Villain" of the piece, no more; and, like the "Villain" in the play, its chief function is to re-act upon the other players for higher ends. There is, in point of fact, a *second* factor which one might venture to call the *Struggle for the Life of Others*, which plays an equally prominent part. Even in the early stages of development, its contribution is as real, while in the world's later progress—under the name of Altruism— it assumes a sovereignty before which the earlier Struggle sinks into insignificance. That this second form of Struggle . should all but have escaped the notice of Evolutionists is the more unaccountable since it arises, like the first, out of those fundamental functions of living organisms which it is the main business of biological science to investigate. The functions discharged by all living things, plant and animal, are two [1] in number. The first is Nutrition, the second is Reproduction. The first is the basis of the Struggle for Life; the second, of the Struggle for the Life of Others. These two functions run their parallel course—or spiral course, for they con- tinuously intertwine—from the very dawn of life. They are involved in the fundamental nature of proto-

---

[1] There is a third function—that of Co-relation—but, to avoid confusing the immediate issue, this may remain at present in the background.

plasm itself. They affect the entire round of life; they determine the whole morphology of living things ; in a sense they are life. Yet, in constructing the fabric of Evolution, one of these has been taken, the other left.

Partly because of the limitations of its purely physical name, and partly because it has never been worked out as an evolutionary force, the function of Reproduction will require to be introduced to the reader in some detail. But to realize its importance or even to understand it, it will be necessary to recall to our minds the supreme place which function generally holds in the economy of life.

Life to an animal or to a Man is not a random series of efforts. Its course is set as rigidly as the courses of the stars. All its movements and changes, its apparent deflections and perturbations are guided by unalterable purposes ; its energies and caprices definitely controlled. What controls it are its functions. These and these only determine life ; living out these is life. Trace back any one, or all, of the countless activities of an animal's life, and it will be found that they are at bottom connected with one or other of the two great functions which manifest themselves in protoplasm. Take any organ of the body—hand or foot, eye or ear, heart or lung—or any tissue of the body—muscle or nerve, bone or cartilage—and it will be found to be connected either with Nutrition or with Reproduction. Just as everything about an engine, every bolt, bar, valve, crank, lever, wheel, has something to do with the work of that engine, everything about an animal's body has something to do with the work prescribed by those two functions. An animal, or a Man, is a consistent whole, a rational production.

Now the rationale of living is revealed for us in protoplasm. Protoplasm sets life its task. Living can only be done along its lines. There start the channels in which all life must run, and though the channels bifurcate endlessly as time goes on, and though more life and fuller is ever coursing through them, it can never overflow the banks appointed from the beginning.

But this is not all. The activities even of the higher life, though not qualitatively limited by the lower, are determined by these same lines. Were these facts only relevant in the domain of physiology, they would be of small account in a study of the Ascent of Man. But the more profoundly the Evolution of Man is investigated the more clearly is it seen that the whole course of his development has been conducted on this fundamental basis. Life, all life, higher or lower, is an organic unity. Nature may vary her effects, may introduce qualitative changes so stupendous as to make their affinities with lower things unthinkable, but she has never re-laid the foundations of the world. Evolution began with protoplasm and ended with Man, and all the way between, the development has been a symmetry whose secret lies in the two or three great crystallizing forces revealed to us through this first basis.

Having realized the significance of the physiological functions, let us now address ourselves to their meaning and connotations. The first, the function of Nutrition, on which the Struggle for Life depends, requires no explanation. Mr. Darwin was careful to give to his favorite phrase, the Struggle for Life, a wider meaning than that which associates it merely with Nutrition; but this qualification seems largely to

have been lost sight of—to some extent even by himself—and the principle as it stands to-day in scientific
and philosophical discussion is practically synonymous with the Struggle for Food. As time goes on
this Struggle—at first a conflict with Nature and the
elements, sustained by hunger, and intensified by
competition—assumes many disguises, and is ultimately known in the modern world under the names
of War and Industry. In these later phases the early
function of protoplasm is obscured, but on the last
analysis, War and Industry—pursuits in which half
the world is now engaged—are seen to be simply its
natural developments.

The implications of the second function, Reproduction, lie further from the surface. To say that Reproduction is synonymous with the Struggle for the Life
of Others conveys at first little meaning, for the
physiological aspects of the function persist in the
mind, and make even a glimpse of its true character
difficult. In two or three chapters in the text, the
implications of this function will be explained at
length, and the reader who is sufficiently interested in
the immediate problem, or who sees that there is here
something to be investigated, may do well to turn to
these at once. Suffice it for the moment to say that
the physiological aspects of the Struggle for the Life
of Others are so overshadowed even towards the close
of the Animal Kingdom by the psychical and ethical
that it is scarcely necessary to emphasize the former
at all. One's first and natural association with the
Struggle for the Life of Others is with something
done for posterity—in the plant the Struggle to produce seeds, in the animal to beget young. But this is

a preliminary which, compared with what directly
and indirectly rises out of it, may be almost passed
over. The significant note is ethical, the development
of Other-ism as Altruism—its immediate and inevitable outcome. Watch any higher animal at that
most critical of all hours—for itself, and for its species
—the hour when it gives birth to another creature
like itself. Pass over the purely physiological processes of birth; observe the behavior of the animal-
mother in presence of the new and helpless life which
palpitates before her. There it lies, trembling in the
balance between life and death. Hunger tortures it;
cold threatens it; danger besets it; its blind existence
hangs by a thread. There is the opportunity of
Evolution. There is an opening appointed in the
physical order for the introduction of a moral order.
If there is more in Nature than the selfish Struggle
for Life the secret can now be told. Hitherto, the
world belonged to the Food-seeker, the Self-seeker, the
Struggler for Life, the Father. Now is the hour of
the Mother. And, animal though she be, she rises to
her task. And that hour, as she ministers to her
young, becomes to her, and to the world, the hour of
its holiest birth.

Sympathy, tenderness, unselfishness, and the long
list of virtues which make up Altruism, are the direct
outcome and essential accompaniment of the reproductive process. Without some rudimentary maternal solicitude for the egg in the humblest forms of
life, or for the young among higher forms, the living
world would not only suffer, but would cease. For a
time in the life-history of every higher animal the
direct, personal, gratuitous, unrewarded help of an-

2

other creature is a condition of existence. Even in the lowliest world of plants the labors of Maternity begin, and the animal kingdom closes with the creation of a class in which this function is perfected to its last conceivable expression. The vicarious principle is shot through and through the whole vast web of Nature; and if one actor has played a mightier part than another in the drama of the past, it has been self-sacrifice. What more has come into humanity along the line of the Struggle for the Life of Others will be shown later. But it is quite certain that, of all the things that minister to the welfare and good of Man, of all that make the world varied and fruitful, of all that make society solid and interesting, of all that make life beautiful and glad and worthy, by far the larger part has reached us through the activities of the Struggle for the Life of Others.

How grave the omission of this supreme factor from our reckoning, how serious the effect upon our whole view of nature, must now appear. Time was when the science of Geology was interpreted exclusively in terms of the action of a single force—fire. Then followed the theories of an opposing school who saw all the earth's formations to be the result of water. Any Biology, any Sociology, any Evolution, which is based on a single factor, is as untrue as the old Geology. It is only when both the Struggle for Life and the Struggle for the Life of Others are kept in view, that any scientific theory of Evolution is possible. Combine them, contrast them, assign each its place, allow for their inter-actions, and the scheme of Nature may be worked out in terms of them to the last detail. All along the line, through the whole course of the development, these two functions act and react upon one another ; and continually as they co-operate, to produce a single result, their specific differences are never lost.

The first, the Struggle for Life, is, throughout, the Self-regarding function ; the second, the Other-regarding function. The first, in lower Nature, obeying the law of self-preservation, devotes its energies to feed itself ; the other, obeying the law of species-preservation, to feed its young. While the first develops the active virtues of strength and courage, the other lays the basis for the passive virtues, sympathy, and love. In the later world one seeks its end in personal aggrandizement, the other in ministration. One begets competition, self-assertion, war; the other unselfishness, self-effacement, peace. One is Individualism, the other, Altruism.

To say that no ethical content can be put into the discharge of either function in the earlier reaches of Nature goes without saying. But the moment we reach a certain height in the development, ethical implications begin to arise. These, in the case of the first, have been read into Nature, lower as well as higher, with an exaggerated and merciless malevolence. The other side has received almost no expression. The final result is a picture of Nature wholly painted in shadow—a picture so dark as to be a challenge to its Maker, an unanswered problem to philosophy, an abiding offence to the moral nature of Man. The world has been held up to us as one great battlefield heaped with the slain, an Inferno of infinite suffering, a slaughter-house resounding with the cries of a ceaseless agony.

Before this version of the tragedy, authenticated by the highest names on the roll of science, humanity was dumb, morality mystified, natural theology stultified. A truer reading may not wholly relieve the first, enlighten the second, or re-instate the third. But it at least re-opens the inquiry; and when all its bearings come to be perceived, the light thrown upon the field of Nature by the second factor may be more impressive to reason than the apparent shadow of the first to sense.

To relieve the strain of the position forced upon ethics by the one-sided treatment of the process of Evolution heroic attempts have been made. Some have attempted to mitigate the amount of suffering it involves, and assure us that, after all, the Struggle, except as a metaphor, scarcely exists. "There is," protests Mr. Alfred Russel Wallace, "good reason to believe that the supposed 'torments' and 'miseries' of animals have little real existence, but are the reflection of the amagined sensations of cultivated men and women in similar circumstances; and that the amount of actual suffering caused by the Struggle for Existence among animals is altogether insignificant." [1] Mr. Huxley, on the other hand, will make no compromise. The Struggle for Life to him is a portentous fact, unmitigated and unexplained. No metaphors are strong enough to describe the implacability of its sway. " The moral indifference of nature " and " the unfathomable injustice of the nature of things " everywhere stare him in the face. " For his successful progress as far as the savage state, Man has been largely indebted to those qualities which he shares with the

[1] *Darwinism,* p. 37.

ape and the tiger."[1] That stage reached, "for thousands and thousands of years, before the origin of the oldest known civilizations, men were savages of a very low type. They strove with their enemies and their competitors; they preyed upon things weaker or less cunning than themselves; they were born, multiplied without stint, and died, for thousands of generations, alongside the mammoth, the urus, the lion, and the hyæna, whose lives were spent in the same way; and they were no more to be praised or blamed, on moral grounds, than their less erect and more hairy compatriots. . . . Life was a continual free fight, and beyond the limited and temporary relations of the family, the Hobbesian war of each against all was the normal state of existence. The human species, like others, plashed and floundered amid the general stream of evolution, keeping its head above water as it best might, and thinking neither of whence nor whither."[2]

How then does Mr. Huxley act—for it is instructive to follow out the consequences of an error—in the face of this tremendous problem? He gives it up. There is no solution. Nature is without excuse. After framing an indictment against it in the severest language at his command, he turns his back upon Nature —sub-human Nature, that is—and leaves teleology to settle the score as best it can. "The history of civilization," he tells us, "is the record of the attempts of the human race to escape from this position." But whither does he betake himself? Is he not part of Nature, and therefore a sharer in its guilt? By no

[1] *Evolution and Ethics*, p. 6.
[2] *Nineteenth Century*, Feb., 1888.

means. For by an astonishing *tour de force*—the last,
as his former associates in the evolutionary ranks
have not failed to remind him, which might have been
expected of him—he ejects himself from the world-
order, and washes his hands of it in the name of Ethical Man. After sharing the fortunes of Evolution all
his life, bearing its burdens and solving its doubts, he
abandons it without a pang, and sets up an *imperium
in imperio,* where, as a moral being, the " cosmic "
Struggle troubles him no more. " Cosmic Nature," he
says, in a parting shot at his former citadel, " is no
school of virtue, but the headquarters of the enemy of
ethical nature." [1]   So far from the Ascent of Man running along the ancient line, " Social progress means a
checking of the cosmic process at every step, and the
substitution for it of another, which may be called the
ethical process ; the end of which is not the survival
of those who may happen to be fittest, in respect of
the whole of the conditions which exist, but of those
who are ethically the best. [2] "

The expedient, to him, was a necessity. Viewing
Nature as Mr. Huxley viewed it there was no other
refuge. The " cosmic process " meant to him the
Struggle for Life, and to escape from the Struggle
for Life he was compelled to turn away from the
world-order, which had its being because of it. As it
happens, Mr. Huxley has hit upon the right solution,
only the method by which he reaches it is wholly
wrong. And the mischievous result of it is obvious
—it leaves all lower Nature in the lurch. With
a curious disregard of the principle of Continuity, to

---

[1] *Evolution and Ethics,* p. 27.   [2] *Ibid.,* p. 33.

which all his previous work had done such homage, he splits up the world-order into two separate halves. The earlier dominated by the "cosmic" principle— the Struggle for Life; the other by the "ethical" principle—virtually, the Struggle for the Life of Others. The Struggle for Life is thus made to stop at the "ethical" process; the Struggle for the Life of Others to begin. Neither is justified by fact. The Struggle for the Life of Others, as we have seen, starts its upward course from the same protoplasm as the Struggle for Life; and the Struggle for Life runs on into the "ethical" sphere as much as the Struggle for the Life of Others. One has only to see where Mr. Huxley gets his "ethical" world to perceive the extent of the anomaly. For where does he get it, and what manner of world is it? "The history of civilization details the steps by which men have succeeded in building up an artificial world within the cosmos." [1] An *artificial* world within the cosmos?

This suggested breach between the earlier and the later process, if indeed we are to take it seriously, is scientifically indefensible, and the more unfortunate since the same result, or a better, can be obtained without it. The real breach is not between the earlier and the later process, but between two rival, or two co-operating processes, which have existed from the first, which have worked together all along the line, and which took on "ethical" characters at the same moment in time. The Struggle for the Life of Others is sunk as deep in the "cosmic process" as the Struggle for Life; the Struggle for Life has a share in the "ethical process" as much as the Strug-

[1] *Evolution and Ethics*, p. 35.

gle for the Life of Others. Both are cosmic processes;
both are ethical processes; both are both cosmical
and ethical processes. Nothing but confusion can
arise from a cross-classification which does justice to
neither half of Nature.

The consternation caused by Mr. Huxley's change
of front, or supposed change of front, is matter of
recent history. Mr. Leslie Stephen and Mr. Herbert
Spencer hastened to protest; the older school of
moralists hailed it almost as a conversion. But the
one fact everywhere apparent throughout the discussion is that neither side apprehended either the
ultimate nature or the true solution of the problem.
The seat of the disorder is the same in both attackers
and attacked—the one-sided view of Nature. Universally Nature, as far as the plant, animal, and
savage levels, is taken to be synonymous with the
Struggle for Life. Darwinism held the monopoly of
that lower region, and Darwinism revenged itself in a
manner which has at least shown the inadequacy of
the most widely-accepted premise of recent science.

That Mr. Huxley has misgivings on the matter
himself is apparent from his Notes. " Of course,"
he remarks, in reference to the technical point,
"strictly speaking, social life and the ethical process
in virtue of which it advances towards perfection are
part and parcel of the general process of Evolution." [1]
And he gets a momentary glimpse of the " ethical
process " in the cosmos, which, if he had followed it
out, must have modified his whole position. " Even
in these rudimentary forms of society, love and fear
come into play, and enforce a greater or less renun-

[1] *Evolution and Ethics*, note 19.

ciation of self-will. To this extent the general cosmic process begins to be checked by a rudimentary ethical process, which is, strictly speaking, part of the former, just as the 'governor' in a steam-engine is part of the mechanism of the engine." [1]

Here the whole position is virtually conceded; and only the pre-conceptions of Darwinism and the lack of a complete investigation into the nature and extent of the "rudimentary ethical process" can have prevailed in the face of such an admission. Follow out the metaphor of the "governor," and, with one important modification, the true situation almost stands disclosed. For what appears to be the "governor" in the rudimentary ethical process becomes the "steam-engine" in the later process. The mere fact that it exists in the "general cosmic process" alters the quality of that process; and the fact that, as we hope to show, it becomes the prime mover in the later process, entirely changes our subsequent conception of it. The beginning of a process is to be read from the end and not from the beginning. And if even a rudiment of a moral order be found in the beginnings of this process it relates itself and that process to a final end and a final unity.

Philosophy reads end into the earlier process by a necessity of reason. But how much stronger its position if it could add to that a basis in the facts of Nature? "I ask the evolutionist," pertinently inquires Mr. Huxley's critic, who has no other basis than the Struggle for existence how he accounts for the intrusion of these moral ideas and standards which presume to interfere with the cosmic process

Evolution and Ethics, note 19.

and sit in judgment upon its results." [1] May we ask
the philosopher how *he* accounts for them? As little
can he account for them as he who has "no other
basis than the Struggle for existence." Truly, the
writer continues, the question "cannot be answered so
long as we regard morality merely as an incidental result, a by-product, as it were, of the cosmical system."
But what if morality be the main product of the cosmical system—of *even* the cosmical system? What if
it can be shown that it is the essential and not the incidental result of it, and that so far from being a by-
product, it is *im*morality that is the by-product?

These interrogations may be too strongly put.
"Accompaniments" of the cosmical system might be
better than "products"; "revelations through that
process" may be nearer the truth than "results" of it.
But what is intended to show is that the moral order
is a continuous line from the beginning, that it has
had throughout, so to speak, a basis in the cosmos,
that upon this, as a trellis-work, it has climbed upwards to the top. The one—the trellis-work—is to be
conceived of as an incarnation; the other—the manifestation—as a revelation; the one is an Evolution
from below, the other an Involution from above.
Philosophy has long since assured us of the last, but
because it was never able to show us the completeness
of the first, science refused to believe it. The de-
faulter nevertheless was not philosophy but science.
Its business was with the trellis-work. And it gave
us a broken trellis-work, a ladder with only one side,
and every step on the other side resting on air. When
science tried to climb the ladder it failed; the steps

[1] Prof. Seth, *Blackwood's Magazine*, Dec., 1893.

refused to bear any weight. What did men of science do? They condemned the ladder and, balancing themselves on the side that was secure, proclaimed their Agnosticism to philosophy. And what did philosophy do? It stood on the other half of the ladder, *the half that was not there*, and rated them. That the other half was not there was of little moment. It was in themselves. It ought to be there; therefore it must be there. And it is quite true; it is there. Philosophy, like Poetry, is prophetic: "The sense of the whole," it says, "comes first." [1]

But science could not accept the alternative. It had looked, and it was not there; from its standpoint the only refuge was Agnosticism—there were no *facts*. Till the facts arrived, therefore, philosophy was powerless to relieve her ally. Science looked to Nature to put in her own ends, and not to philosophy to put them in for her. Philosophy might interpret them after they were there, but it must have something to start from; and all that science had supplied her with meantime was the fact of the Struggle for Life. Working from the stand-point of the larger Nature, Human Nature itself, philosophy could put in other ends; but there appeared no solid backing for these in facts, and science refused to be satisfied. The position was a fair one. The danger of philosophy putting in the ends is that she cannot convince every one that they are the right ones.

And what is the valid answer? Of course, that Nature has put in her own ends if we would take the trouble to look for them. She does not require them to be secretly manufactured upstairs and credited to

[1] Prof. H. Jones, *Browning*, p. 28.

her account. By that process mistakes might arise
in the reckoning. The philosophers upstairs might
differ about the figures, or at least in equating them.
The philosopher requires fact, phenomenon, natural
law, at every turn to keep him right; and without at
least some glimpse of these, he may travel far afield.
So long as Schopenhauer sees one thing in the course
of Nature and Rousseau another, it will always be
well to have Nature herself to act as referee. The
end as read in Nature and the end as re-read in, and
interpreted by, the higher Nature of Man may be very
different things ; but nothing can be done till the End-
in-the-phenomenon clears the way for the End-in-
itself—till science overtakes philosophy with facts.
When that is done, everything can be done. With
the finding of the other half of the ladder, even Agnosticism may retire. Science cannot permanently
pronounce itself " not knowing," till it has exhausted
the possibilities of knowing. And in this case the
Agnosticism is premature, for science has only to look
again, and it will discover that the missing facts are
there.

Seldom has there been an instance on so large a
scale of a biological error corrupting a whole philosophy. Bacon's aphorism was never more true:
" This I dare affirm in knowledge of Nature, that a
little natural philosophy, and the first entrance into
it, doth dispose the opinion to atheism, but on the
other side, much natural philosophy, and wading deep
into it, will bring about men's minds to religion." [1]
Hitherto, the Evolutionist has had practically no other
basis than the Struggle for Life. Suppose even we

[1] *Meditationes Sacræ,* X.

leave that untouched, the addition of an Other-regarding basis makes an infinite difference. For when it is then asked on which of them the process turns, and the answer is given "On both," we perceive that it is neither by the one alone, nor by the other alone, that the process is to be interpreted, but by a higher unity which resolves and embraces all. And as both are equally necessary to this antinomy, even that of the two which seems irreconcilable with higher ends is seen to be necessary. Viewed *simpliciter*, the Struggle for Life appears irreconcilable with ethical ends, a prodigious anomaly in a moral world ; but viewed in continuous reaction with the Struggle for the Life of Others, it discloses itself as an instrument of perfection the most subtle and far-reaching that reason could devise.

The presence of the second factor, therefore, while it leaves the first untouched, cannot leave its implications untouched. It completely alters these implications. It has never been denied that the Struggle for Life is an efficient instrument of progress ; the sole difficulty has always been to justify the nature of the instrument. But if even it be shown that this is only half the instrument, teleology gains something. If the fuller view takes nothing away from the process of Evolution, it imports something into it which changes the whole aspect of the case. For even from the first that factor is there. The Struggle for the Life of Others, as we have seen, is no interpolation at the end of the process, but radical, engrained in the world-order as profoundly as the Struggle for Life. By what right, then, has Nature been interpreted only by the Struggle for Life? With far greater justice might

science interpret it in the light of the Struggle for the
Life of Others. For, in the first place, unless there
had been this second factor, the world could not have
existed. Without the Struggle for the Life of Others,
obviously there would have been no Others. In the
second place, unless there had been a Struggle for the
Life of Others, the Struggle for Life could not have
been kept up. As will be shown later the Struggle
for Life almost wholly supports itself on the products
of the Struggle for the Life of Others. In the third
place, without the Struggle for the Life of Others, the
Struggle for Life as regards its energies would have
died down, and failed of its whole achievement. It is
the ceaseless pressure produced by the exuberant fertility of Reproduction that creates any valuable Struggle for Life at all. The moment "Others" multiply,
the individual struggle becomes keen up to the dis-
ciplinary point. It was this, indeed—through the
reading of Malthus on Over-population—that suggested to Mr. Darwin the value of the Struggle for
Life. The law of Over-population from that time forward became the foundation-stone of his theory; and
recent biological research has made the basis more
solid than ever. The Struggle for the Life of Others
on the plant and animal plane, in the mere work of
multiplying lives, is a final condition of progress.
Without competition there can be no fight, and without fight there can be no victory. In other words,
without the Struggle for the Life of Others there can
be no Struggle for Life, and therefore no Evolution.
Finally, and all the reasons already given are frivolous
beside it, had there been no Altruism—Altruism in
the definite sense of unselfishness, sympathy, and self-

sacrifice for Others, the whole higher world of life had perished as soon as it was created. For hours, or days, or weeks in the early infancy of all higher animals, maternal care and sympathy are a condition of existence. Altruism *had* to enter the world, and any species which neglected it was extinguished in a generation.

No doubt a case could be made out likewise for the imperative value of the Struggle for Life. The position has just been granted. So far from disputing it, we assume it to be equally essential to Nature and to a judgment upon the process of Evolution. But what is disputed is that the Struggle for Life is either the key to Nature, or that it is more important in itself than the Struggle for the Life of Others. It is pitiful work pitting the right hand against the left, the heart against the head; but if it be insisted that there is neither right hand nor heart, the proclamation is necessary not only that they exist, but that absolutely they are as important and relatively to ethical Man of infinitely greater moment than anything that functions either in the animal or social organism.

But why, if all this be true of the Struggle for the Life of Others, has a claim so imperious not been recognized by science? That a phenomenon of this distinction should have attracted so little attention suggests a suspicion. Does it really exist? Is the biological basis sound? Have we not at least exaggerated its significance? The biologist will judge. Though no doubt the function of Reproduction is intimately connected in Physiology with the function of Nutrition, the facts as stated here are facts of Nature; and some glimpse of the influence of

this second factor will be given in the sequel from
which even the non-biological reader may draw his
own conclusions. Difficult as it seems to account for
the ignoring of an elemental fact in framing the
doctrine of Evolution, there are circumstances which
make the omission less unintelligible. Foremost, of
course, there stands the overpowering influence of
Mr. Darwin. In spite of the fact that he warned his
followers against it, this largely prejudged the issue.
Next is to be considered the narrowing, one had almost said the blighting, effect of specialism. Necessary to the progress of science, the first era of a reign
of specialism is disastrous to philosophy. The men
who in field and laboratory are working out the facts,
do not speculate at all. Content with slowly building
up the sum of actual knowledge in some neglected and
restricted province, they are too absorbed to notice
even what the workers in the other provinces are
about. Thus it happens that while there are many
scientific men, there are few scientific thinkers. The
complaint is often made that science speculates too
much. It is quite the other way. One has only to
read the average book of science in almost any department to wonder at the wealth of knowledge, the
brilliancy of observation, and the barrenness of idea.
On the other hand, though scientific experts will not
think themselves, there is always a multitude of on-
lookers waiting to do it for them. Among these what
strikes one is the ignorance of fact and the audacity of
the idea. The moment any great half-truth in Nature
is unearthed, these unqualified practitioners leap to a
generalization ; and the observers meantime, on the
track of the other half, are too busy or too oblivious to

refute their heresies. Hence, long after its foundations are undermined, a brilliant generalization will retain its hold upon the popular mind ; and before the complementary, the qualifying, or the neutralizing facts can be supplied, the mischief is done.

But while this is true of many who play with the double-edged tools of science, it is not true of a third class. When we turn to the pages of the few whose science is adequate and whose sweep is over the whole vast horizon, we find, as we should expect, some recognition of the altruistic factor. Though Mr. Herbert Spencer, to whom the appeal in this connection is obvious, makes a different use of the fact, it has not escaped him. Not only does the Other-regarding function receive recognition, but he allots it a high place in his system. Of its ethical bearings he is equally clear. " What," he asks, "is the ethical aspect of these altruistic principles ? In the first place, animal life of all but the lowest kinds has been maintained by virtue of them. Excluding the *Protozoa*, among which their operation is scarcely discernible, we see that without *gratis* benefits to offspring, and earned benefits to adults, life could not have continued. In the second place, by virtue of them life has gradually evolved into higher forms. By care of offspring which has become greater with advancing organization, and by survival of the fittest in the competition among adults, which has become more habitual with advancing organization, superiority has been perpetually fostered and further advances caused." [1] Fiske, Littré, Romanes, Le Conte, L. Büchner, Miss Buckley, and Prince Kropotkin have expressed them-

[1] *Principles of Ethics,* Vol. II., p. 5.

selves partly in the same direction; and Geddes and
Thomson, in so many words, recognize "the co-exist-
ence of twin-streams of egoism and altruism, which
often merge for a space without losing their distinct-
ness, and are traceable to a common origin in the
simplest forms of life." [1] The last named—doubtless
because their studies have taken them both into the
fields of pure biology and of bionomics—more clearly
than any other modern writers, have grasped the
bearings of this theme in all directions, and they fearlessly take their stand-point from the physiology of
protoplasm. Thus, "in the hunger and reproductive
attractions of the lowest organisms, the self-regarding
and other-regarding activities of the higher find their
starting-point. Though some vague consciousness is
perhaps co-existent with life itself, we can only speak
with confidence of psychical egoism and altruism
after a central nervous system has been definitely established. At the same time, the activities of even the
lowest organisms are often distinctly referable to
either category. . . . Hardly distinguishable at
the outset, the primitive hunger and love become the
starting-points of divergent lines of egoistic and altruistic emotion and activity." [2]

That at a much earlier stage than is usually supposed, Evolution *visibly* enters upon the "rudimentary ethical" plane, is certain, and we shall hope to
outline the proof. But even if the thesis fails, it remains to challenge the general view that the Struggle
for Life is everything, and the Struggle for the Life
of Others nothing. Seeing not only that the second is
the more important; but also this far more significant

[1] *The Evolution of Sex*, p. 279.   [2] *Ibid.*, p. 279.

fact—which has not yet been alluded to—that *as Evolution proceeas the one Struggle waxes, and the other wanes,* would it not be wiser to study the drama nearer its *dénouement* before deciding whether it was a moral, a non-moral, or an immoral play ?

Lest the alleged *waning* of the Struggle for Life convey a wrong impression, let it be added that of course the word is to be taken qualitatively. The Struggle in itself can never cease. What ceases is its so-called anti-ethical character. For nothing is in finer evidence as we rise in the scale of life than the gradual tempering of the Struggle for Life. Its slow amelioration is the work of ages, may be the work of ages still, but its animal qualities in the social life of Man are being surely left behind ; and though the mark of the savage and the brute still mar its handiwork, these harsher qualities must pass away. In that new social order which the gathering might of the altruistic spirit is creating now around us, in that reign of Love which must one day, if the course of Evolution holds on its way, be realized, the baser elements will find that solvent prepared for them from the beginning in anticipation of a higher rule on earth. Interpreting the course of Evolution scientifically, whether from its starting-point in the first protoplasm, or from the rallying-point of its two great forces in the social organism of to-day, it becomes more and more certain that only from the commingled achievement of both can the nature of the process be truly judged. Yet, as one sees the one sun set, and the other rise with a splendor the more astonishing and bewildering as the centuries roll on, it is impossible to withhold a verdict as to which may be most reasonably looked upon as the ultimate reality of the world. The path of progress and the path of Altruism are one. Evolution is nothing but the Involution of Love, the revelation of Infinite Spirit, the Eternal Life returning to Itself. Even the great shadow of Egoism which darkens the past is revealed as shadow only because we are compelled to read it by the higher light which has come. In the very act of judging it to be shadow, we assume and vindicate the light. And in every vision of the light, contrariwise, we resolve the shadow, and perceive the end for which both light and dark are given.

> " I can believe, this dread machinery
> Of sin and sorrow would confound me else.
> Devised—all pain, at most expenditure
> Of pain by Who devised pain—to evolve,
> By new machinery in counterpart,
> The moral qualities of Man—how else ?—
> To make him love in turn, and be beloved,
> Creative and self-sacrificing too,
> And thus eventually Godlike." [1]

III.

WHY WAS EVOLUTION THE METHOD CHOSEN.

ONE seldom-raised yet not merely curious question of Evolution is, why the process should be an evolution at all? If Evolution, is simply a method of Creation, why was this very extraordinary method chosen? Creation *tout d'un coup* might have produced the same result; an instantaneous act or an age-long process

The Ring and the Book—The Pope, 1375.

would both have given us the world as it is? The answer of modern natural theology has been that the evolutionary method is the infinitely nobler scheme. A spectacular act, it is said, savors of the magician. As a mere exhibition of power it appeals to the lower nature; but a process of growth suggests to the reason the work of an intelligent Mind. No doubt this intellectual gain is real. While a catastrophe puts the universe to confusion at the start, a gradual rise makes the beginning of Nature harmonious with its end. How the surpassing grandeur of the new conception has filled the imagination and kindled to enthusiasm the soberest scientific minds, from Darwin downwards, is known to every one. As the memorable words which close the *Origin of Species* recall: "There is a grandeur in this view of life, with its several powers, having been originally breathed by the Creator into a few forms or into one; and that whilst this planet has gone cycling on, according to the fixed law of gravity, from so simple a beginning endless forms most beautiful and most wonderful have been, and are being evolved." [1]

But can an intellectual answer satisfy us any more than the mechanical answer which it replaced? As there was clearly a moral purpose in the end to be achieved by Evolution, should we not expect to find some similar purpose in the means? Can we perceive no high design in selecting this particular design, no worthy ethical result which should justify the conception as well as the execution of Evolution?

We go too far, perhaps, in expecting answers to questions so transcendent. But one at least suggests

[1] *Origin of Species*, p. 429.

itself, whose practical value is apology enough for
venturing to advance it. Whenever the scheme was
planned, it must have been foreseen that the time
would come when the directing of part of the course
of Evolution would pass into the hands of Man. A
spectator of the drama for ages, too ignorant to see
that it was a drama, and too impotent to do more than
play his little part, the discovery must sooner or later
break upon him that Nature meant him to become a
partner in her task, and share the responsibility of the
closing acts. It is not given to him as yet to bind the
sweet influences of Pleiades, or to unloose the bands of
Orion. In part only can he make the winds and waves
obey him, or control the falling rain. But in larger
part he holds the dominion of the world of lower life.
He exterminates what he pleases; he creates and he
destroys; he changes; he evolves; his selection re-
places natural selection; he replenishes the earth with
plants and animals according to his will. But in a far
grander sphere, and in an infinitely profounder sense,
has the sovereignty passed to him. For, by the same
decree, he finds himself the guardian and the arbiter
of his personal destiny, and that of his fellow-men.
The moulding of his life and of his children's children
in measure lie with him. Through institutions of his
creation, through Parliaments, Churches, Societies,
Schools, he shapes the path of progress for his country
and his time. The evils of the world are combated by
his remedies; its passions are stayed, its wrongs re-
dressed, its energies for good or evil directed by his
hand. For unnumbered millions he opens or shuts the
gates of happiness, and paves the way for misery or
social health. Never before was it known and felt

with the same solemn certainty that Man, within bounds which none can pass, must be his own maker and the maker of the world. For the first time in history not individuals only but multitudes of the wisest and the noblest in every land take home to themselves, and unceasingly concern themselves with the problem of the Evolution of Mankind. ⸱ Multitudes more, philanthropists, statesmen, missionaries, humble men and patient women, devote themselves daily to its practical solution, and everywhere some, in a God-like culmination of Altruism, give their very lives for their fellow-men. Who is to help these Practical Evolutionists—for those who read the book of Nature can call them by no other name, and those who know its spirit can call them by no higher—who is to help them in their tremendous task? There is the will—where is the wisdom?

Where but in Nature herself. Nature may have entrusted the further building to Mankind, but the plan has never left her hands. The lines of the future are to be learned from her past, and her fellow-helpers can most easily, most loyally, and most perfectly do their part by studying closely the architecture of the earlier world, and continuing the half-finished structure symmetrically to the top. The information necessary to complete the work with architectural consistency lies in Nature. We might expect that it should be there. When a business is transferred, or a partner assumed, the books are shown, the methods of the business explained, its future developments pointed out. All this is now done for the Evolution of Mankind. In Evolution Creation has shown her hand. To have kept the secret from Man would have imperilled the further evolution. To have revealed it sooner had been premature. Love must come before knowledge, for knowledge is the instrument of Love, and useless till it arrives. But now that there is Altruism enough in the world to begin the new era, there must be wisdom enough to direct it. To make Nature spell out her own career, to embody the key to the development in the very development itself, so that the key might be handed over along with the work, was to make the transference of responsibility possible and rational. In the seventeenth century, Descartes, who with Leibnitz already foresaw the adumbration of the evolutionary process, almost pointed this out; for speaking, in another connection, of the intellectual value of a slow development of things he observes, " their nature is much more easy to conceive when they are seen originating by degrees in this way, than when they are considered as entirely made." [1]

The past of Nature is a working-model of how worlds can be made. The probabilities are there is no better way of making them. If Man does as well it will be enough. In any case he can only begin where Nature left off, and work with such tools as are put into his hands. If the new partner had been intended merely to experiment with world-making, no such legacy of useful law had been ever given him. And if he had been meant to begin *de novo* on a totally different plan, it is unlikely either that that should not have been hinted at, or that in his touching and beautiful endeavor he should be embarrassed and thrown off the track by the old plan. As a child set to complete some fine embroidery is shown the stitches, the

[1] *Discourse on Method.*

colors, and the outline traced upon the canvas, so the great Mother in setting their difficult task to her later children provides them with one superb part finished to show the pattern.

IV.

EVOLUTION AND SOCIOLOGY.

The moment it is grasped that we may have in Nature a key to the future progress of Mankind, the study of Evolution rises to an imposing rank in human interest. There lies the programme of the world from the first of time, the instrument, the charter, and still more the prophecy of progress. Evolution is the natural directory of the sociologist, the guide through that which has worked in the past to what—subject to modifying influences which Nature can always be trusted to give full notice of—may be expected to work in the future. Here, for the individual, is a new and impressive summons to public action, a vocation chosen of Nature which it will profit him to consider, for thereby he may not only save the whole world, but find his own soul. "The study of the historical development of man," says Prof. Edward Caird, "especially in respect of his higher life, is not only a matter of external or merely speculative curiosity; it is closely connected with the development of that life in ourselves. For we learn to know ourselves, first of all, in the mirror of the world: or, in other words, our knowledge of our own nature and of its possibilities grows and deepens with

our understanding of what is without us, and most of
all with our understanding of the general history of
man. It has often been noticed that there is a certain
analogy between the life of the individual and that of
the race, and even that the life of the individual is a
sort of epitome of the history of humanity. But, as
Plato already discovered, it is by reading the large
letters that we learn to interpret the small.   .   .   .
It is only through a deepened consciousness of the
world that the human spirit can solve its own problem. Especially is this true in the region of anthropology. For the inner life of the individual is deep
and full just in proportion to the width of his relations
to other men and things; and his consciousness of
what he is in himself as a spiritual being is dependent
on a comprehension of the position of his individual
life in the great secular process by which the intellectual and moral life of humanity has grown and is
growing. Hence the highest practical, as well as speculative, interests of men are connected with the new
extension of science which has given fresh interest and
meaning to the 'whole history of the race." [1] If, as
Herbert Spencer reminds us, " it is one of those open
secrets which seem the more secret because they are
so open, that all phenomena displayed by a nation are
phenomena of Life, and are dependent on the laws of
Life," we cannot devote ourselves to study those laws
too earnestly or too soon. From the failure to get at
the heart of the first principles of Evolution the old
call to " follow Nature " has all but become a heresy.
Nature, as a moral teacher, thanks to the Darwinian
interpretation, was never more discredited than at

[1] *The Evolution of Religion*, Vol. i., pp. 25, 29.

this hour; and friend and foe alike agree in warning
us against her. But a further reading of Nature may
decide not that we must discharge the teacher but beg
her mutinous pupils to try another term at school.
With Nature studied in the light of a true biology, or
even in the sense in which the Stoics themselves employed their favorite phrase, it must become once
more the watchword of personal and social progress.
With Mr. Huxley's definition of what the Stoics
meant by Nature as " that which holds up the ideal of
the supreme good and demands absolute submission of
the will to its behests. . . which commands all men
to love one another, to return good for evil, to regard
one another as citizens of one great state," [1] the
phrase, "Live according to Nature," so far from having no application to the modern world or no sanction in modern thought, is the first commandment of
Natural Religion.

The sociologist has grievously complained of late
that he could get but little help from science. The
suggestions of Bagehot, the Synthetic Philosophy of
Herbert Spencer, the proposals of multitudes of the
followers of the last who announced the redemption
of the world the moment they discovered the " Social
Organisms," raised great expectations. But somehow
they were not fulfilled. Mr. Spencer's work has been
mainly to give this century, and in part all time, its
first great map of the field. He has brought all the
pieces on the board, described them one by one, defined and explained the game. But what he has
failed to do with sufficient precision, is to pick out the
King and Queen. And because he has not done so,

[1] *Evolution and Ethics*, p. 27.

some men have mistaken his pawns for kings; others
have mistaken the real kings for pawns; every *ism*
has found endorsement in his pages, and men have
gathered courage for projects as hostile to his whole
philosophy as to social order. Theories of progress
have arisen without any knowledge of its laws, and
the ordered course of things has been done violence to
by experiments which, unless the infinite conservatism of Nature had neutralized their evils, had been
a worse disaster than they are. This inadequacy, indeed, of modern sociology to meet the practical problems of our time, has become a by-word. Mr. Leslie
Stephen pronounces the existing science "a heap of
vague empirical observation, too flimsy to be useful";
and Mr. Huxley, exasperated with the condition in
which it leaves the human family, prays that if
"there is no hope of a large improvement" he should
"hail the advent of some kindly comet which would
sweep the whole affair away."

The first step in the reconstruction of Sociology will
be to escape from the shadow of Darwinism—or rather
to complement the Darwinian formula of the Struggle
for Life by a second factor which will turn its darkness into light. A new morphology can only come
from a new physiology, and *vice versa;* and for both
we must return to Nature. The one-sided induction
has led Sociology into a wilderness of empiricism, and
only a complete induction can reinstate it among the
sciences. The vacant place is there awaiting it; and
every earnest mind is prepared to welcome it, not only
as the coming science, but as the crowning Science of
all the sciences, the Science, indeed, for which it will
one day be seen every other science exists. What it

waits for meantime is what every science has had to wait for, exhaustive observation of the facts and ways of Nature. Geology stood still for centuries waiting for those who would simply look at the facts. Men speculated in fantastic ways as to how the world could have been made, and the last thing that occurred to them was to go and see it making. Then came the observers, men who, waiving all theories of the process, addressed themselves to the natural world direct, and in watching its daily programme of falling rain and running stream laid bare the secret for all time. Sociology has had its Werners; it awaits its Huttons. The method of Sociology must be the method of all the natural sciences. It also must go and see the world making, not where the conditions are already abnormal beyond recall, or where Man, by irregular action, has already obscured everything but the conditions of failure; but in lower Nature which makes no mistakes, and in those fairer reaches of a higher world where the quality and the stability of the progress are guarantees that the eternal order of Nature has had her uncorrupted way.

It cannot be that the full programme for the perfect world lies in the imperfect part. Nor can it ever be that science can find the end in the beginning, get moral out of non-moral states, evolve human societies from ant-heaps, or philanthropies from protoplasm. But in every beginning we get a beginning of an end; in every process a key to the single step to be taken next. The full corn is not in the ear, but the first cell of it is, and though "it doth not yet appear" what the million-celled ear shall be, there is rational ground for judging what the second cell shall be. The next

few cells of the Social Organism are all that are given
to Sociology to affect. And, in dealing with them, its
business is with the forces; the phenomena will
take care of themselves. Neither the great forces
of Nature, nor the great lines of Nature, change in a
day, and however apparently unrelated seem the phenomena as we ascend—here animal, there human; at
one time non-moral, at another moral—the lines of
progress are the same. Nature, in *horizontal* section,
is broken up into strata which present to the eye of
ethical Man the profoundest distinctions in the universe; but Nature in the *vertical* section offers no
break, or pause, or flaw. To study the first is to study
a hundred unrelated sciences, sciences of atoms, sciences of cells, sciences of Souls, sciences of Societies;
to study the second is to deal with one science—Evolution. Here, on the horizontal section, may be what
Geology calls an unconformability; there is overlap;
changes of climate may be registered from time to
time each with its appropriate re-action on the things
contained; upheavals, depressions, denudations, glacia-
tions, faults, vary the scene; higher forms of fossils
appear as we ascend; but the laws of life are continuous throughout, the eternal elements in an ever
temporal world. The Struggle for Life, and the
Struggle for the Life of Others, in essential nature
have never changed. They find new expression in
each further sphere, become colored to our eye with
different hues, are there the rivalries or the affections
of the brute, and here the industrial or the moral
conflicts of the race; but the factors themselves remain the same, and all life moves in widening spirals
round them. Fix in the mind this distinction between

the horizontal and the vertical view of Nature, between the phenomena and the law, between all the sciences that ever were and the one science which resolves them all, and the confusions and contradictions of Evolution are reconciled. The man who deals with Nature statically, who catalogues the phenomena of life and mind, puts on each its museum label, and arranges them in their separate cases, may well defy you to co-relate such diverse wholes. To him Evolution is alike impossible and unthinkable. But these items that he labels are not wholes. And the world he dissects is not a museum, but a living, moving and ascending thing. The sociologist's business is with the vertical section, and he who has to do with this living, moving, and ascending thing must treat it from the dynamic point of view.

The significant thing for him is the study of Evolution on its working side. And he will find that nearly all the phenomena of social and national life are phenomena of these two principles—the Struggle for Life, and the Struggle for the Life of Others. Hence he must betake himself in earnest to see what these mean in Nature, what gathers round them as they ascend, how each acts separately, how they work together, and whither they seem to lead. More than ever the method of Sociology must be biological. More urgently than ever "the time has come for a better understanding and for a more radical method; for the social sciences to strengthen themselves by sending their roots deep into the soil underneath from which they spring, and for the biologist to advance over the frontier and carry the methods of his science boldly into human society, where he has but to deal with the phenomena of life, where he encounters life at last under its highest and most complex aspect." [1]

Would that the brilliant writer whose words these are, and whose striking work appears while these sheets are almost in the press, had " sent his roots deep enough into biological soil " to discover the true foundation for that future Science of Society which he sees to be so imperative. No modern thinker has seen the problem so clearly as Mr. Kidd, but his solution, profoundly true in itself, is vitiated in the eyes of science and philosophy by a basis wholly unsound. With an emphasis which Darwin himself has not excelled, he proclaims the enduring value of the Struggle for Life. He sees its immense significance even in the highest ranges of the social sphere. There it stands with its imperious call to individual assertion, inciting to a rivalry which Nature herself has justified, and encouraging every man by the highest sanctions ceaselessly to seek his own. But he sees nothing else in Nature; and he encounters therefore the difficulty inevitable from this stand-point. For to obey this voice means ruin to Society, wrong and anarchy against the higher Man. He listens for another voice; but there is no response. As a social being he cannot, in spite of Nature, act on his first initiative. He must subordinate himself to the larger interest, present and future, of those around him. But why, he asks, *must* he, since Nature says " Mind thyself?" Till Nature adds the further precept, " Look not every man on his own things, but also on the things of Others," there is no rational sanction for morality. And he finds no such

[1] Benjamin Kidd, *Social Evolution*, p. 28.

precept. There is none in Nature. There is none in Reason. Nature can only point him to a strenuous rivalry as the one condition of continued progress; Reason can only endorse the verdict. Hence he breaks at once with reason and with Nature, and seeks an "ultra-rational sanction" for the future course of social progress.

Here, in his own words, is the situation. "The teaching of reason to the individual must always be that the present time and his own interests therein are all-important to him. That the forces which are working out our development are primarily concerned not with those interests of the individual, but with those widely different interests of a social organism subject to quite other conditions and possessed of an indefinitely longer life. . . . The central fact with which we are confronted in our progressive societies is, therefore, that the interests of the social organism and those of the individuals comprising it at any time are actually antagonistic; they can never be reconciled; they are inherently and essentially irreconcilable." [1] Observe the extraordinary dilemma. Reason not only has no help for the further progress of Society, but Society can only go on upon a principle which is an affront to it. As Man can only attain his highest development in Society, his individual interests must more and more subordinate themselves to the welfare of a wider whole. "How is the possession of reason ever to be rendered compatible with the will to submit the conditions of existence so onerous, requiring the effective and continual subordination of the individual's welfare to the progress of a develop-

[1] *Op. cit.*, p. 78.

4

ment in which he can have no personal interest whatever ? " [1]

Mr. Kidd's answer is the bold one that it is not compatible. There is no rational sanction whatever for
progress. Progress, in fact, can only go on by enlist-
ing Man's reason against itself. "All those systems
of moral philosophy, which have sought to find in the
nature of things a rational sanction for human conduct
in society, must sweep round and round in futile
circles. They attempt an inherently impossible task.
The first great social lesson of those evolutionary doc-
trines which have transformed the science of the nineteenth century is, that there cannot be such a sanction.[2] . . . The extraordinary character of the
problem presented by human society begins thus
slowly to come into view. We find man making continual progress upwards, progress which it is almost
beyond the power of the imagination to grasp. From
being a competitor of the brutes he has reached a
point of development at which he cannot himself set
any limits to the possibilities of further progress, and
at which he is evidently marching onwards to a high
destiny. He has made this advance under the sternest conditions, involving rivalry and competition for
all, and the failure and suffering of great numbers.
His reason has been, and necessarily continues to be, a
leading factor in this development; yet, granting, as
we apparently must grant, the possibility of the re-
versal of the conditions from which his progress
results, those conditions have not any sanction from
his reason. They have had no such sanction at any
stage of his history, and they continue to be as much

[1] *Op. cit.*, p. 64.   [2] *Op. cit.*, p. 79.

without such sanction in the highest civilization of the
present day as at any past period." [1]

These conclusions will not have been quoted in vain
if they show the impossible positions to which a
writer, whose contribution otherwise is of profound
and permanent value, is committed by a false reading
of Nature. Is it conceivable, *a priori*, that the human
reason should be put to confusion by a breach of the
Law of Continuity at the very point where its sustained action is of vital moment? The whole complaint, which runs like a dirge through every chapter
of this book, is founded on a misapprehension of the
fundamental laws which govern the processes of
Evolution. The factors of Darwin and Weismann
are assumed to contain an ultimate interpretation of
the course of things. For all time the conditions of
existence are taken as established by these authorities.
With the Struggle for Life in sole possession of the
field no one, therefore, we are warned, need ever
repeat the gratuitous experiment of the past, of
Socrates, Plato, Kant, Hegel, Comte, and Herbert
Spencer, to find a sanction for morality in Nature.
"All methods and systems alike, which have endeavored to find in the nature of things any universal
rational sanction for individual conduct in a progressive society, must be ultimately fruitless. They are
all alike inherently unscientific in that they attempt
to do what the fundamental conditions of existence
render impossible." And Mr. Kidd puts a climax on
his devotion to the doctrine of his masters by mourn-
ing over "the incalculable loss to English Science
and English Philosophy" because Herbert Spencer's

[1] *Op. cit.*, pp. 77–78.

work "was practically complete before his intellect had any opportunity of realizing the full transforming effect in the higher regions of thought, and, more particularly, in the department of sociology, of that development of biological science which began with Darwin, which is still in full progress, and to which Professor Weismann has recently made the most notable contributions." [1] Whether Mr. Spencer's ignorance or his science has been at the bottom of the escape, it is at least a lucky one. For if Mr. Kidd had realized "the full transforming effect" of the following paragraph, much of his book could not have been written. "The most general conclusion is that in order of obligation, the preservation of the species takes precedence of the preservation of the individual. It is true that the species has no existence save as an aggregate of individuals; and it is true that, therefore, the welfare of the species is an end to be subserved only as subserving the welfare of individuals. But since disappearance of the species, implying absolute disappearance of all individuals, involves absolute failure in achieving the end, whereas disappearance of individuals though carried to a great extent, may leave outstanding such numbers as can, by continuance of the species, make subsequent fulfilment of the end possible; the preservation of the individual must, in a variable degree according to circumstances, be subordinated to the preservation of the species, where the two conflict." [2]

What Mr. Kidd has succeeded, and splendidly succeeded, in doing is to show that Nature *as interpreted in terms of the Struggle for Life* contains no

[1] *Op. cit.*, p. 80   [2] *Principles of Ethics*, Vol. II., p. 6.

sanction either for morality or for social progress. But instead of giving up Nature and Reason at this point, he should have given up Darwin. The Struggle for Life is *not* " the supreme fact up to which biology has slowly advanced." It is the fact to which Darwin advanced; but if biology had been thoroughly consulted it could not have given so maimed an account of itself. With the final conclusion reached by Mr. Kidd we have no quarrel. Eliminate the errors due to an unrevised acceptance of Mr. Darwin's interpretation of Nature, and his work remains the most important contribution to Social Evolution which the last decade has seen. But what startles us is his method. To put the future of Social Science on an ultra-rational basis is practically to give it up. Unless thinking men have some sense of the consistency of a method they cannot work with it, and if there is no guarantee of the stability of the results it would not be worth while.

But all that Mr. Kidd desires is really to be found in Nature. There is no single element even of his highest sanction which is not provided for in a thorough-going doctrine of Evolution—a doctrine, that is, which includes all the facts and all the factors, and especially which takes into account that evolution of Environment which goes on *pari passu* with the evolution of the organism and where the highest sanctions ultimately lie. With an Environment which widens and enriches until it includes—or consciously includes, for it has never been absent—the Divine; and with Man so evolving as to become more and more conscious that that Divine is there, and above all that it is in himself, all the materials and all the sanctions for a moral progress are forever secure. None of the sanctions of religion are withdrawn by adding to them the sanctions of Nature. Even those sanctions which are supposed to lie over and above Nature may be none the less rational sanctions. Though a positive religion, in the Comtian sense, is no religion, a religion that is not in some degree positive is an impossibility. And although religion must always rest upon faith, there is a reason for faith, and a reason not only in Reason, but in Nature herself. When Evolution comes to be worked out along its great natural lines, it may be found to provide for all that religion assumes, all that philosophy requires, and all that science proves.

Theological minds, with premature approval, have hailed Mr. Kidd's solution as a vindication of their supreme position. Practically, as a vindication of the dynamic power of the religious factor in the Evolution of Mankind, nothing could be more convincing. But as an apologetic, it only accentuates a weakness which scientific theology never felt more keenly than at the present hour. This weakness can never be removed by an appeal to the ultra-rational. Does Mr. Kidd not perceive that any one possessed of reason enough to encounter his dilemma, either in the sphere of thought or of conduct, will also have reason enough to reject any " ultra-rational" solution ? This dilemma is not one which would occur to more than one in a thousand ; it has tasked all Mr. Kidd's powers to convince his reader that it exists ; but if exceptional intellect is required to see it, surely exceptional intellect must perceive that this is not the way out of it. One cannot, in fact, *think* oneself out of a difficulty of this kind ; it can only be *lived* out. And that precisely is what Nature is making all of us, in greater or less degree, do, and every day making us do more. By the time, indeed, that the world as a whole is sufficiently educated to see the problem, it will already have been solved. There is little comfort, then, for apologetics in this direction. Only by bringing theology into harmony with Nature and into line with the rest of our knowledge can the noble interests given it to conserve retain their vitality in a scientific age. The first essential of a working religion is that it shall be congruous with Man ; the second that it shall be congruous with Nature. Whatever its sanctions, its forces must not be abnormal, but reinforcements and higher potentialities of those forces which, from eternity, have shaped the progress of the world. No other dynamic can enter into the working schemes of those who seek to guide the destinies of nations or carry on the Evolution of Society on scientific principles. A divorce here would be the catastrophe of reason, and the end of faith. We believe with Mr. Kidd that " the process of social development which has been taking place, and which is still in progress, in our Western civilization, is not the product of the intellect, but the motive force behind it has had its seat and origin in the *fund of altruistic feeling* with which our civilization has become equipped." But we shall endeavor to show that this fund of altruistic feeling has been slowly funded in the race by Nature, or through Nature, and as the direct and inevitable result of that Struggle for the Life of Others, which has been from all time a condition of existence. What religion has done to build up this fund, it may not be within the scope of this introductory volume to inquire; it has done so much that students of religion may almost be pardoned the oversight of the stupendous natural basis which made it possible. But nothing is gained by protesting that "this altruistic development, and the deepening and softening of character which has accompanied it, are the *direct and peculiar product* of the religious system." For nothing can ever be gained by setting one half of Nature against the other, or the rational against the ultra-rational. To affirm that Altruism is a peculiar product of religion is to excommunicate Nature from the moral order, and religion from the rational order. If science is to begin to recognize religion, religion must at least end by recognizing science. And so far from religion sacrificing vital distinctions by allying itself with Nature, so far from impoverishing its immortal quality by accepting some contribution from the lower sphere, it thereby extends itself over the whole rich field, and claims all—matter, life, mind, space, time—for itself. The present danger is not in applying Evolution as a method, but only in not carrying it far enough. No man, no man of science even, observing the simple facts, can ever rob religion of its due. Religion has done more for the development of Altruism in a few centuries than all the millenniums of geological time. But we dare not rob Nature of its due. We dare not say that Nature played the prodigal for ages, and reformed at the eleventh hour. If Nature is the Garment of God, it is woven without seam throughout; if a revelation of God, it is the same yesterday, to-day, and forever; if the expression of His Will, there is in it no variable-ness nor shadow of turning. Those who see great gulfs fixed—and we have all begun by seeing them—end by seeing them filled up. Were these gulfs essential to any theory of the universe or of Man, even the establishment of the unity of Nature were a dear price to pay for obliterating them. But the apparent loss is only gain, and the seeming gain were infinite loss. For to break up Nature is to break up Reason, and with it God and Man.

CHAPTER I.

THE ASCENT OF THE BODY.

THE earliest home of Primitive Man was a cave in the rocks—the simplest and most unevolved form of human habitation. One day, perhaps driven by the want within his hunting-grounds of the natural cave, he made himself a hut—an artificial cave. This simple dwelling-place was a one-roomed hut or tent of skin and boughs, and so completely does it satisfy the rude man's needs that down to the present hour no ordinary savage improves upon the idea. But as the hut surrounds itself with other huts and grows into a village, a new departure must take place. The village must have its chief, and the chief, in virtue of his larger life, requires a more spacious home. Each village, therefore, adds to its one-roomed hut, a hut with two rooms. From the two-roomed hut we pass, among certain tribes, to three- and four-roomed huts, and finally to the many-chambered lodge of the Head-Chief or King.

This passage from the simple cave to the many-chambered lodge is an Evolution, and a similar development may be traced in the domestic architecture of all civilized societies. The laborer's cottage of mod-

ern England and the shieling of the Highland crofter
are the survivals of the one-roomed hut of Primitive
Man, scarcely changed in any essential with the lapse
of years. In the squire's mansion also, and the noble-
man's castle, we have the representatives, but now in
an immensely developed form, of the many-roomed
home of the chief. The steps by which the cottage
became the castle are the same as those by which the
cave in the rocks became the lodge of the chief.
Both processes wear the hall-mark of all true development—they arise in response to growing necessities, and they are carried out by the most simple and
natural steps.

In this evolution of a human habitation we have an
almost perfect type of the evolution of that more
august habitation, the complex tenement of clay in
which Man's mysterious being has its home. The
Body of Man is a structure of a million, or a million
million cells. And the history of the unborn babe is,
in the first instance, a history of additions, of room
being added to room, of organ to organ, of faculty to
faculty. The general process, also, by which this
takes place is almost as clear to modern science as in
the case of material buildings. A special class of observers has carefully watched these secret and amazing metamorphoses, and so wonderful has been their
success with mind and microscope that they can almost claim to have seen Man's Body made. The Science of Embryology undertakes to trace the development of Man from a stage in which he lived in a one-
roomed house—a physiological cell. Whatever the
multitude of rooms, the millions and millions of cells,
in which to-day each adult carries on the varied work

of life, it is certain that when he first began to be he was the simple tenant of a single cell. Observe, it is not some animal-ancestor or some human progenitor of Man that lived in this single cell—that may or may not have been—but the individual Man, the present occupant himself. We are dealing now not with phylogeny—the history of the race—but with ontogeny—the problem of Man's Ascent from his own earlier self. And the point at the moment is not that the race ascends; it is that each individual man has once, in his own life-time, occupied a single cell, and starting from that humble cradle, has passed through stage after stage of differentiation, increase, and development, until the myriad-roomed adult-form was attained. Whence that first cradle came is at present no matter. Whether . its remote progenitor rocked among the waves of primeval seas or swung from the boughs of forests long since metamorphosed into coal does not affect the question of the individual ascent of Man. The answers to these questions are hypotheses. The *fact* that now arrests our wonder is that when the earliest trace of an infant's organization meets the eye of science it is nothing but a one-celled animal. And so closely does its development from that distant point follow the lines of the evolution just described in the case of the primitive savage hut, that we have but to make a few changes in phraseology to make the one process describe the other. Instead of rooms and chambers we shall now read cells and tissues; instead of the builder's device of adding room to room, we shall use the physiologist's term *segmentation;* the employments carried on in the various rooms will become the functions discharged by the organs of the

human frame, and line for line the history of the evolution will be found to be the same.

The embryo of the future man begins life, like the
primitive savage, in a one-roomed hut, a single simple
cell.   This cell is round and almost microscopic in
size.   When fully formed it measures only one-tenth
of a line in diameter, and with the naked eye can be
barely discerned as a very fine point.   An outer covering, transparent as glass, surrounds this little sphere,
and in the interior, embedded in protoplasm, lies a
bright globular spot.   In form, in size, in composition
there is no apparent difference between this human
cell and that of any other mammal.   The dog, the elephant, the lion, the ape, and a thousand others begin
their widely different lives in a house the same as
Man's.   At an earlier stage indeed, before it has taken
on its pellucid covering, this cell has affinities still
more astonishing.   For at that remoter period the earlier forms of all living things, both plant and animal,
are one.   It is one of the most astounding facts of
modern science that the first embryonic abodes of
moss and fern and pine, of shark and crab and coral
polyp, of lizard, leopard, monkey, and Man are so
exactly similar that the highest powers of mind and
microscope fail to trace the smallest distinction between them.

But let us watch the development of this one-celled
human embryo.   Increase of rooms in architecture can
be effected in either of two ways—by building entirely
new rooms, or by partitioning old ones.   Both of these
methods are employed in Nature.   The first, gemma-
tion, or budding, is common among the lower forms of
life.   The second, differentiation by partition, or seg-

mentation, is the approved method among higher animals, and is that adopted in the case of Man. It proceeds, after the fertilized ovum has completed the complex preliminaries of karyokinesis, by the division of the interior-contents into two equal parts, so that the original cell is now occupied by two nucleated cells with the old cell-wall surrounding them outside. The two-roomed house is, in the next development, and by a similar process of segmentation, developed into a structure of four rooms, and this into one of eight, and so on.[1] In a short time the number of chambers is so

[1] When the multicellular globe, made up of countless offshoots or divisions of the original pair, has reached a certain size, its centre becomes filled with a tiny lakelet of watery fluid. This fluid gradually increases in quantity and, pushing the cells outward, packs them into a single layer, circumscribing it on every side as with an elastic wall. At one part a dimple soon appears, which slowly deepens, until a complete hollow is formed. So far does this invagination of the sphere go on that the cells at the bottom of the hollow touch those at the opposite side. The ovum has now become an open bag or cup, such as one might make by doubling in an india-rubber ball, and thus is formed the *gastrula* of biology. The evolutional interest of this process lies in the fact that probably all animals above the Protozoa pass through this gastrula stage. That some of the lower *Metazoa*, indeed, never develop much beyond it, a glance at the structure of the humbler Coelenterates will show—the simplest of all illustrations of the fact that embryonic forms of higher animals are often permanently represented by the adult forms of lower. The chief thing however to mark here is the doubling-in of the ovum to gain a double instead of a single wall of cells. For these two different layers, the ectoderm and the endoderm, or the animal layer and the vegetal layer, play a unique part in the after-history. All the organs of movement and sensation spring from the one, all the organs of nutrition and reproduction develop from the other.

great that count is lost, and the activity becomes so
vigorous in every direction that one ceases to notice
individual cells at all. The tenement in fact consists
now of innumerable groups of cells congregated together, suites of apartments as it were, which have
quickly arranged themselves in symmetrical, definite,
and withal different forms. Were these forms not
different as well as definite we should hardly call it an
evolution, nor should we characterize the resulting
aggregation as a higher organism. A hundred cottages placed in a row would never form a castle.
What makes the castle superior to the hundred cottages is not the number of its rooms, for they are possibly fewer; nor their difference in shape, for that is
immaterial. It lies in the number and nature and
variety of useful purposes to which the rooms are put,
the perfection with which each is adapted to its end,
and the harmonious co-operation among them with
reference to some common work. This also is the distinction between a higher animal and a humble organism such as the centipede or the worm. These
creatures are a monotony of similar rings, like a string
of beads. Each bead is the counterpart of the other;
and with such an organization any high or varied life
becomes an impossibility. The fact that any growing
embryo is passing through a real development is decided by the new complexity of structure, by the more
perfect division of labor, and of better kinds of labor,
and by the increase in range and efficiency of the correlated functions discharged by the whole. In the
development of the human embryo the differentiating
and integrating forces are steadily acting and co-operating from the first, so that the result is not a mere

aggregation of similar cells, but an organism with different parts and many varied functions. When all is complete we find that one suite of cells has been especially set apart to provide the commissariat, others have devoted themselves exclusively to assimi- lation. The ventilation of the house—respiration— has been attended to by others, and a central force- pump has been set up, and pipes and ducts for many purposes installed throughout the system. Telegraph wires have next been stretched in every direction to keep up connection between the endless parts; and other cells developed into bony pillars for support. Finally, the whole delicate structure has been shielded by a variety of protective coverings, and after months and years of further elaboration and adjustment the elaborate fabric is complete. Now all these com- plicated contrivances—bones, muscles, nerves, heart, brain, lungs—are made out of cells; they are them- selves, and in their furthest development, simply masses or suites of cells modified in various ways for the special department of household work they are meant to serve. No new thing, except building material, has entered into the embryo since its first appearing. It seized whatever matter lay to hand, incorporated it with its own quickening substance, and built it in to its appropriate place. So the structure rose in size and symmetry, till the whole had climbed, a miracle of unfolding, to the stature of a Man.

But the beauty of this development is not the sig- nificant thing to the student of Evolution; nor is it the occultness of the process nor the perfection of the result that fill him with awe as he surveys the finished work. It is the immense distance Man has come.

Between the early cell and the infant's formed body, the ordinary observer sees the uneventful passage of a few brief months. But the evolutionist sees concentrated into these few months the labor and the progress of incalculable ages. Here before him is the whole stretch of time since life first dawned upon the earth; and as he watches the nascent organism climbing to its maturity he witnesses a spectacle which for strangeness and majesty stands alone in the field of biological research. What he sees is not the mere shaping or sculpturing of a Man. The human form does not begin as a human form. It begins as an animal; and at first, and for a long time to come there is nothing wearing the remotest semblance of humanity. What meets the eye is a vast procession of lower forms of life, a succession of strange inhuman creatures emerging from a crowd of still stranger and still more inhuman creatures; and it is only after a prolonged and unrecognizable series of metamorphoses that they culminate in some faint likeness to the image of him who is one of the newest yet the oldest of created things. Hitherto we have been taught to look among the fossiliferous formations of Geology for the buried lives of the earth's past. But Embryology has startled the world by declaring that the ancient life of the earth is not dead. It is risen. It exists to-day in the embryos of still-living things, and some of the most archaic types find again a resurrection and a life in the frame of man himself.

It is an amazing and almost incredible story. The proposition is not only that Man begins his earthly existence in the guise of a lower animal-embryo, but that in the successive transformations of the human

embryo there is reproduced before our eyes a visible,
actual, physical representation of part of the life-
history of the world. Human Embryology is a condensed account, a recapitulation or epitome of some of
the main chapters in the Natural History of the world.
The same processes of development which once
took thousands of years for their consummation are
here condensed, foreshortened, concentrated into the
space of weeks. Each platform reached by the human
embryo in its upward course represents the embryo of
some lower animal which in some mysterious way has
played a part in the pedigree of the human race, which
may itself have disappeared long since from the earth,
but is now and forever built into the inmost being of
Man. These lower animals, each at its successive
stage, have stopped short in their development ; Man
has gone on. At each fresh advance his embryo is
found again abreast of some other animal-embryo a
little higher in organization than that just passed.
Continuing his ascent that also is overtaken, the now
very complex embryo making up to one animal-embryo after another until it has distanced all in its series
and stands alone. As the modern stem-winding watch
contains the old clepsydra and all the most useful
features in all the timekeepers that were ever made ;
as the Walter printing-press contains the rude hand-
machine of Gutenberg, and all the best in all the
machines that followed it ; as the modern locomotive
of to-day contains the engine of Watt, the locomotive
of Hedley, and most of the improvements of succeeding
years, so Man contains the embryonic bodies of earlier
and humbler and clumsier forms of life. Yet in
making the Walter press in a modern workshop, the

artificer does not begin by building again the press of
Gutenberg, nor in constructing the locomotive does
the engineer first make a Watt's machine and then
incorporate the Hedley, and then the Stephenson, and
so on through all the improving types of engines that
have led up to this. But the astonishing thing is that,
in making a Man, Nature does introduce the framework of these earlier types, displaying each crude
pattern by itself before incorporating it in the finished
work. The human embryo, to change the figure, is a
subtle phantasmagoria, a living theatre in which a
weird transformation scene is being enacted, and in
which countless strange and uncouth characters take
part. Some of these characters are well-known to
science, some are strangers. As the embryo unfolds,
one by one these animal actors come upon the stage,
file past in phantom-like procession, throw off their
drapery, and dissolve away into something else. Yet,
as they vanish, each leaves behind a vital portion of
itself, some original and characteristic memorial, something itself has made or won, that perhaps it alone
could make or win—a bone, a muscle, a ganglion, or a
tooth—to be the inheritance of the race. And it is
only after nearly all have played their part and dedi-
cated their gift, that a human form, mysteriously
compounded of all that has gone before, begins to be
discerned in their midst.

The duration of this process, the profound antiquity
of the last survivor, the tremendous height he has
scaled, are inconceivable by the faculties of Man. But
measure the very lowest of the successive platforms
passed in the ascent, and see how very great a thing
it is even to rise at all. The single cell, the first

definite stage which the human embryo attains, is still
the adult form of countless millions both of animals
and plants. Just as in modern England the millionaire's mansion—the evolved form—is surrounded by
laborers' cottages—the simple form—so in Nature,
living side by side with the many-celled higher animals, is an immense democracy of unicellular artizans.
These simple cells are perfect living things. The
earth, the water, and the air teem with them everywhere. They move, they eat, they reproduce their
like. But one thing they do not do—they do not rise.
These organisms have, as it were, stopped short in the
ascent of life. And long as evolution has worked
upon the earth, the vast numerical majority of plants
and animals are still at this low stage of being. So
minute are some of these forms that if their one-
roomed huts were arranged in a row it would take
twelve thousand to form a street a single inch in
length. In their watery cities—for most of them are
Lake-Dwellers—a population of eight hundred thousand million could be accommodated within a cubic
inch. Yet, as there was a period in human history
when none but cave-dwellers lived in Europe, so was
there a time when the highest forms of life upon the
globe were these microscopic things. See, therefore,
the meaning of Evolution from the want of it. In a
single hour or second the human embryo attains the
platform which represents the whole life-achievement
of myriads of generations of created things, and the
next day or hour is immeasurable centuries beyond
them.

Through all what zoological regions the embryo
passes in its great ascent from the one-celled forms,

one can never completely tell. The changes succeed
one another with such rapidity that it is impossible
at each separate stage, to catch the actual likeness
to other embryos. Sometimes a familiar feature suddenly recalls a form well-known to science, but the
likeness fades, and the developing embryo seems to
wander among the ghosts of departed types. Long
ago these crude ancestral forms were again the highest animals upon the earth. For a few thousand
years they reigned supreme, furthered the universal
evolution by a hair-breadth, and passed away. The
material dust of their bodies is laid long since in the
Palæozoic rocks, but their life and labor are not
forgotten. For their gains were handed on to a succeeding race. Transmitted thence through an endless
series of descendants, sifted, enriched, accentuated,
still dimly recognizable, they re-appeared at last in
the physical frame of Man. After the early stages of
human development are passed, the transformations
become so definite that the features of the contributory animals are almost recognizable. Here, for
example, is a stage at which the embryo in its anatomical characteristics resembles that of the Vermes
or Worms. As yet there is no head, nor neck, nor
backbone, nor waist, nor limbs. A roughly cylindri-
cal headless trunk—that is all that stands for the
future man. One by one the higher Invertebrates are
left behind, and then occurs the most remarkable
change in the whole life-history. This is the laying
down of the line to be occupied by the spinal chord,
the presence of which henceforth will determine the
place of Man in the Vertebrate sub-kingdom. At this
crisis, the eye which sweeps the field of lower Nature

for an analogue will readily find it. It is a circumstance of extraordinary interest that there should be
living upon the globe at this moment an animal
representing the actual transition from Invertebrate
to Vertebrate life. The acquisition of a vertebral
column is one of the great marks of height which
Nature has bestowed upon her creatures ; and in the
shallow waters of the Mediterranean she has preserved for us a creature which, whether degenerate
or not, can only be likened to one of her first rude
experiments in this direction. This animal is the
Lancelet, or Amphioxus, and so rudimentary is the
backbone that it does not contain any bone at all, but
only a shadow or prophecy of it in cartilage. The
cartilaginous *notochord* of the Amphioxus nevertheless
is the progenitor of all vertebral columns, and in the
first instance this structure appears in the human
embryo exactly as it now exists in the Lancelet. But
this is only a single example. In living Nature there
are a hundred other animal characteristics which at
one stage or another the biologist may discern in the
ever-changing kaleidoscope of the human embryo.

Even with this addition, nevertheless, the human
infant is but a first rough draft, an almost formless
lump of clay. As yet there is no distinct head, no
brain, no jaws, no limbs; the heart is imperfect, the
higher visceral organs are feebly developed, everything is elementary. But gradually new organs loom
in sight, old ones increase in complexity. By a magic
which has never yet been fathomed the hidden Potter
shapes and re-shapes the clay. The whole grows in
size and symmetry. Resemblances, this time, to
the embryos of the lower vertebrate series, flash out as

each new step is attained—first the semblance of the
Fish, then of the Amphibian, then of the Reptile, last
of the Mammal. Of these great groups the leading
embryonic characters appear as in a moving panorama, some of them pronounced and unmistakable,
others mere sketches, suggestions, likenesses of infinite
subtlety. At last the true Mammalian form emerges
from the crowd. Far ahead of all at this stage stand
out three species—the Tailed Catarrhine Ape, the Tail-
less Catarrhine, and last, differing physically from
these mainly by an enlargement of the brain and a
development of the larynx, Man.

Whatever views be held of the doctrine of Evolution, whatever theories of its cause, these facts of
Embryology are proved. They have taken their place
in science wholly apart from the discussion of theories
of Evolution, and as the result of laboratory investigation, made for quite other ends. What is true for
Man, moreover, is true of all other animals. Every
creature that lives climbs up its own genealogical tree
before it reaches its mature condition. "All animals
living, or that ever have lived, are united together by
blood relationship of varying nearness or remoteness,
and every animal now in existence has a pedigree
stretching back, not merely for ten or a hundred
generations, but through all geologic time since life
first commenced on the earth. The study of development has revealed to us that each animal bears the
mark of its ancestry, and is compelled to discover its
parentage in its own development; the phases through
which an animal passes in its progress from the egg to
the adult are no accidental freaks, no mere matters of
developmental convenience, but represent more or less

closely, in more or less modified manner, the successive ancestral stages through which the present condition has been acquired." [1] Almost foreseen by Agassiz, suggested by Von Baer, and finally applied by Fritz Müller, this singular law is the key-note of modern Embryology. In no case, it is true, is the recapitulation of the past complete. Ancestral stages are constantly omitted, others are over-accentuated, condensed, distorted, or confused; while new and undecipherable characters occasionally appear. But it is a general scientific fact, that over the graves of a myriad aspirants the bodies of Man and of all higher Animals have risen. No one knows why this should be so. Science, at present, has no rationale of the process adequate to explain it. It was formerly held that the entire animal creation had contributed something to the anatomy of Man; or that as Serres expressed it, " Human Organogenesis is a transitory Comparative Anatomy." But though Man has not such a monopoly of the past as is here inferred— other types having here and there diverged and developed along lines of their own—it is certain that the materials for his body have been brought together from an unknown multitude of lowlier forms of life.

Those who know the Cathedral of St. Mark's will remember how this noblest of the Stones of Venice owes its greatness to the patient hands of centuries and centuries of workers, how every quarter of the globe has been spoiled of its treasures to dignify this single shrine. But he who ponders over the more ancient temple of the Human Body will find imagination fail

[1] Marshall, *Vertebrate Embryology,* p. 26.

him as he tries to think from what remote and mingled sources, from what lands, seas, climates, atmos-
pheres, its various parts have been called together,
and by what innumerable contributory creatures,
swimming, creeping, flying, climbing, each of its
several members was wrought and perfected. What
ancient chisel first sculptured the rounded columns of
the limbs ? What dead hands built the cupola of the
brain, and from what older ruins were the scattered
pieces of its mosaic-work brought ? Who fixed the
windows in its upper walls ? What winds and
weathers wrought strength into its buttresses ? What
ocean-beds and forest glades worked up its colorings ?
What Love and Terror and Night called forth the
Music ? And what Life and Death and Pain and
Struggle put all together in the noiseless workshop of
the past, and removed each worker silently when its
task was done ? How these things came to be Biology
is one long record. The architects and builders of
this mighty temple are not anonymous. Their names,
and the work they did, are graven forever on the walls
and arches of the Human Embryo. For this is a
volume of that Book in which Man's members were
written, which in continuance were fashioned, when as
yet there was none of them.

The Descent of Man from the Animal Kingdom is
sometimes spoken of as a degradation. It is an un-
speakable exaltation. Recall the vast antiquity of
that primal cell from which the human embryo first
sets forth. Compass the nature of the potentialities
stored up in its plastic substance. Watch all the
busy processes, the multiplying energies, the mystify-
ing transitions, the inexplicable chemistry of this liv-

ing laboratory. Observe the variety and intricacy of its metamorphoses, the exquisite gradation of its ascent, the unerring aim with which the one type unfolds—never pausing, never uncertain of its direction, refusing arrest at intermediate forms, passing on to its flawless maturity without waste or effort or fatigue. See the sense of motion at every turn, of purpose and of aspiration. Discover how, with identity of process and loyalty to the type, a hair-breadth of deviation is yet secured to each so that no two forms come out the same, but each arises an original creation, with features, characteristics, and individualities of its own. Remember, finally, that even to make the first cell possible, stellar space required to be swept of matter, suns must needs be broken up, and planets cool, the agents of geology labor millennium after millennium at the unfinished earth to prepare a material resting-place for the coming guest. Consider all this, and judge if Creation could have a sublimer meaning, or the Human Race possess a more splendid genesis.

From the lips of the Prophet another version, an old and beautiful story, was told to the childhood of the earth, of how God made Man ; how with His own hands He gathered the Bactrian dust, modelled it, breathed upon it, and it became a living soul. Later, the insight of the Hebrew Poet taught Man a deeper lesson. He saw that there was more in Creation than mechanical production. He saw that the Creator had different kinds of Hands and different ways of modelling. How it was done he knew not, but it was not the surface thing his forefathers taught him. The higher divinity and mystery of the process broke upon him. Man was a fearful and wonderful thing. He was modelled in secret. He was curiously wrought in the lowest parts of the earth. When Science came, it was not to contradict the older versions. It but gave them content and a still richer meaning. What the Prophet said, and the Poet saw, and Science proved, all and equally will abide forever. For all alike are voices of the Unseen, commissioned to different peoples and for different ends to declare the mystery of the Ascent of Man.

CHAPTER II.

THE SCAFFOLDING LEFT IN THE BODY.

THE spectacle which we have just witnessed is invisible, and therefore more or less unimpressive, except to the man of science. Embryology works in the dark. Requiring not only the microscope, but the comparative knowledge of intricate and inaccessible forms of life, its all but final contribution to the theory of Evolution carries no adequate conviction to the general mind. We must therefore follow the fortunes of the Body further into the open day. If the Embryo in every changing feature of its growth contains some reminiscence of an animal ancestry, the succeeding stages of its development may be trusted to carry on the proof. And though here the evidence is neither so beautiful nor so exact, we shall find that there is in the adult frame, and even in the very life and movement of the new-born babe, a continuous witness to the ancient animal strain.

We are met, unfortunately, at the outset by one of those curious obstacles to inquiry which have so often barred the way of truth and turned discovery into ridicule. It happens that the class of animals in which Science, in the very nature of the case, is compelled to look for the closest affinities to human beings is that of the *Apes*. This simple circumstance has told almost fatally against the wide acceptance of the theory of Descent. There is just as much truth in the sarcasm that man is a " reformed monkey " as to pre- judge the question to the unscientific mind. But the statement is no nearer the truth itself than if one were to say that a gun is an adult form of the pistol. The connection, if any, between Man and Ape is simply that the most Man-like thing in creation is the Ape, and that, in his Ascent, Man probably passed through a stage when he more nearly resembled the Ape than any other known animal. Apart from that accident, Evolution owes no more to the Ape than to any other creature. Man and Ape are alike in being two of the latest terms of an infinite series, each member of which has had a share in making up the genealogical tree. To single out the Ape, therefore, and use the hypothetical relationship for rhetorical purposes is, to say the least, unscientific. It is certainly the fact that Man is not descended from any existing Ape. The Anthropoid Apes branched off laterally at a vastly remote period from the nearest human progen- itors. The challenge even to produce links between Man and the living man-like Apes is difficult to take seriously. Should any one so violate the first princi- ples of Evolution as to make it, it is only to be said that it cannot be met. For an Anthropoid Ape could as little develop into a Man as could a Man pass back- wards into an Anthropoid Ape. References to a Sim- ian stem play no necessary part in the story of the Ascent of Man. In those pages the compromising name will scarcely occur. If historical sequence compels us to make an apparent exception here at the very outset, it will be seen that the allusion is harmless. For the analogy we are about to make might with equal relevancy have been drawn from a squirrel or a sloth.

On the theory that human beings were once allied in habit as well as in body with some of the Apes, that they probably lived in trees, and that baby-men clung to their climbing mothers as baby-monkeys do to-day, Dr. Louis Robinson prophesied that a baby's power of grip might be found to be comparable in strength to that of a young monkey at the same period of development. Having special facilities for such an investigation, he tested a large number of just-born infants with reference to this particular. Now although most people have some time or other been seized in the awful grasp of a baby, few have any idea of the abnormal power locked up in the tentacles of this human octopus. Dr. Robinson's method was to extend to infants, generally of one hour old, his finger, or a walking stick, to imitate the branch of a tree, and see how long they would hang there without, what the newspapers call, "any other visible means of support." The results are startling. Dr. Robinson has records of upwards of sixty cases in which the children were under a month old, and in at least half of these the experiment was tried within an hour of birth: "In every instance, with only two exceptions, the child was able to hang on to the finger or a small stick, three-quarters of an inch in diameter, by its hands, like an acrobat from a horizontal bar, and sustain the whole weight of its body for at least ten seconds. In twelve cases, in infants under an hour old, half a minute passed before the grasp relaxed, and in three or four nearly a minute. When about four days old, I found that the strength had increased, and that nearly all, when tried at this age, could sustain their weight for half a minute. About a fortnight or three weeks after birth the faculty appeared to have attained its maximum, for several at this period succeeded in hanging for over a minute and a half, two for just over two minutes, and one infant of three weeks old for two minutes thirty-five seconds. . . . In one instance, in which the performer had less than one hour's experience of life, he hung by both hands to my forefinger for ten seconds, and then deliberately let go with his right hand (as if to seek a better hold), and maintained his position for five seconds more by the left hand only. Invariably the thighs are bent nearly at right angles to the body, and in no case did the lower limbs hang down and take the attitude of the erect position. This attitude, and the disproportionately large development of the arms compared with the legs, give the photographs a striking resemblance to a well-known picture of the celebrated Chimpanzee Sally at the Zoological Garden. I think it will be acknowledged that the remarkable strength shown in the flexor muscle of the fore-arm in these young infants, especially when compared with the flaccid and feeble state of the muscular system generally, is a sufficiently striking phenomenon to provoke inquiry as to its cause and origin. The fact that a three-week old baby can perform a feat of muscular strength that would tax the powers of many a healthy adult is enough to set one wondering. A curious point is that in many cases no sign of distress is evident, and no cry uttered until the grasp begins to give way." [1]

Place side by side with this the following account, which Mr. Wallace gives us in his *Malay Archipelago*, of a baby Orang-outang, whose mother he happened to shoot :

" This little creature was only about a foot long, and had, evidently been hanging to its mother when she first fell. Luckily it did not appear to have been wounded, and after we had cleaned the mud out of its mouth it began to cry out, and seemed quite strong and active. While carrying it home it got its hands in my beard, and grasped so tightly that I had great difficulty in getting free, for the fingers are habitually bent inward at the last joint so as to form complete hooks. For the first few days it clung desperately with all four hands to whatever it could lay hold of, and I had to be careful to keep my beard out of its way, as its fingers clutched hold of hair more tenaciously than anything else, and it was impossible to free myself without assistance. When restless, it would struggle about with its hands up in the air trying to find something to take hold of, and when it had got a bit of stick or rag in two or three of its hands, seemed quite happy. For want of something else, it would often seize its own feet, and after a time it would constantly cross its arms and grasp with each hand the long hair that grew just below the opposite shoulder. The great tenacity of its grasp soon diminished, and I was obliged to invent some means to give it exercise and strengthen its limbs. For this purpose

---

[1] *Nineteenth Century*, November, 1891.

I made a short ladder of three or four rounds, on which I put it to hang for a quarter of an hour at a time. At first it seemed much pleased, but it could not get all four hands in a comfortable position, and, after changing about several times, would leave hold of one hand after the other and drop on to the floor. Sometimes when hanging only by two hands, it would loose one, and cross it to the opposite shoulder, grasping its own hair; and, as this seemed much more agreeable than the stick, it would then loose the other and tumble down, when it would cross both and lie on its back quite contentedly, never seeming to be hurt by its numerous tumbles. Finding it so fond of hair, I endeavored to make an artificial mother, by wrapping up a piece of buffalo-skin into a bundle, and suspending it about a foot from the floor. At first this seemed to suit it admirably, as it could sprawl its legs about and always find some hair, which it grasped with the greatest tenacity." [1]

Whatever the value of these facts as evidence, they form an interesting if slight introduction to the part of the subject that lies before us. For we have now to explore the Body itself for actual betrayals—not mere external movements which might have come as well from early Man as from later animal; but veritable physical survivals, the material scaffolding itself—of the animal past. And the facts here are as numerous and as easily grasped as they are authentic. As the traveller, wandering in foreign lands, brings back all manner of curios to remind him where he has been—clubs and spears, clothes and pottery, which

[1] *Malay Archipelago*, 53–5.

represent the ways of life of those whom he has met—
so the body of Man, emerging from its age-long journey through the animal kingdom, appears laden with
the spoils of its distant pilgrimage. These relics are
not mere curiosities ; they are as real as the clubs and
spears, the clothes and pottery. Like them, they were
once a part of life's vicissitude ; they represent organs
which have been outgrown ; old forms of apparatus
long since exchanged for better, yet somehow not yet
destroyed by the hand of time. The physical body of
Man, so great is the number of these relics, is an old
curiosity shop, a museum of obsolete anatomies, dis-
carded tools, outgrown and aborted organs. All other
animals also contain among their useful organs a
proportion which are long past their work ; and so
significant are these rudiments of a former state of
things, that anatomists have often expressed their
willingness to stake the theory of Evolution upon their
presence alone.

Prominent among these vestigial structures, as they
are called, are those which smack of the sea. If Embryology is any guide to the past, nothing is more
certain than that the ancient progenitors of Man once
lived an aquatic life. At one time there was nothing
else in the world but water-life ; all the land animals
are late inventions. One reason why animals began
in the water is that it is easier to live in the water—
anatomically and physiologically cheaper—than to live
on the land. The denser element supports the body
better, demanding a less supply of muscle and bone ;
and the perpetual motion of the sea brings the food to
the animal, making it unnecessary for the animal to
move to the food. This and other correlated circum-

stances calls for far less mechanism in the body, and, as a matter of fact, all the simplest forms of life at the present day are inhabitants of the water.

A successful attempt at coming ashore may be seen in the common worm. The worm is still so unacclimatized to land life that instead of living on the earth like other creatures, it lives *in* it, as if it were a thicker water, and always where there is enough moisture to keep up the traditions of its past. Probably it took to the shore originally by exchanging, first the water for the ooze at the bottom, then by wriggling among muddy flats when the tide was out, and finally, as the struggle for life grew keen, it pushed further and further inland, continuing its migration so long as dampness was to be found.

More striking examples are found among the molluscs, the sea-faring animals *par excellence* of the past. A snail wandering over the earth with a sea-shell on its back is one of the most anomalous sights in nature —as preposterous as the spectacle of a Red Indian perambulating Paris with a birch canoe on his head. The snail not only carries this relic of the sea everywhere with it, but when it cannot get moisture to remind it of its ancient habitat, it actually manufactures it. That the creature itself has discovered the anomaly of its shell is obvious, for in almost every class its state of dilapidation betrays that its up-keep is no longer an object of much importance. In nearly every species the stony houses have already lost their doors, and most have their shells so reduced in size that not half of the body can get in. The degeneration in their cousins, the slugs, is even more pathetic. All that remains of the ancestral home in the highest ranks is a limpet-like cap on the tip of the tail; the lowest are *sans* everything; and in the intermediate forms the former glory is ironically suggested by a few grains of sand or a tiny shield so buried beneath the skin that only the naturalist's eye can see it.

When Man left the water, however—or what was to develop into Man—he took very much more ashore with him than a shell. Instead of crawling ashore at the worm stage, he remained in the water until he evolved into something like a fish; so that when, after an amphibian interlude, he finally left it, many "ancient and fish-like" characters remained in his body to tell the tale. The chief characteristic of a fish is its apparatus for breathing the air dissolved in the water. This consists of gills—delicate curtains hung on strong arches and dyed scarlet with the blood which continually courses through them. In many fishes these arches are five or seven in number, and communicating with them—in order to allow the aerated water, which has been taken in at the mouth, to pass out again after bathing the gills—an equal number of slits or openings is provided in the neck. Sometimes the slits are bare and open so that they are easily seen on the fish's neck—any one who looks at a shark will see them—but in modern forms they are generally covered by the *operculum* or lid. Without these holes in their neck all fishes would instantly perish, and we may be sure Nature took exceptional care in perfecting this particular piece of the mechanism.

Now it is one of the most extraordinary facts in natural history that these slits in the fish's neck are still represented in the neck of Man. Almost the most prominent feature, indeed, after the head, in every mammalian embryo, are the four clefts or furrows of the old gill-slits. They are still known in Embryology by the old name—gill-slits—and so persistent are these characters that children are known to have been born with them not only externally visible—which is a common occurrence—but open through and through, so that fluids taken in at the mouth could pass through and trickle out at the neck. This last fact was so astounding as to be for a long time denied. It was thought that, when this happened, the orifice must have been accidentally made by the probe of the surgeon. But Dr. Sutton has recently met with actual cases where this has occurred. " I have seen milk," he says, " issue from such fistulæ in individuals who have never been submitted to sounding." [1] In the common case of children born with these vestiges, the old gill-slits are represented by small openings in the skin on the sides of the neck, and capable of admitting a thin probe. Sometimes even the place where they have been in childhood is marked throughout life by small round patches of white skin.

Almost more astonishing than the fact of their persistence is the use to which Nature afterwards put them.   When the fish came ashore, its water-breathing apparatus was no longer of any use to it.   At first it had to keep it on, for it took a long time to perfect the air-breathing apparatus destined to replace it. But when this was ready the problem arose, What was to be done with the earlier organ?   Nature is

[1] *Evolution and Disease,* p. 81.

exceedingly economical, and could not throw all this mechanism away. In fact, Nature almost never parts with any structure she has once made. What she does is to change it into something else. Conversely, Nature seldom makes anything new; her method of creation is to adapt something old. Now, when Nature had done with the old breathing-apparatus, she proceeded to adapt it for a new and important purpose. She saw that if water could pass through a hole in the neck, air could pass through likewise. But it was no longer necessary that air should pass through for purposes of *breathing*, for that was already provided for by the mouth. Was there any other purpose for which it was desirable that air should enter the body ? There was, and a very subtle one. For *hearing.* Sound is the result of a wave-motion conducted by many things, but in a special way by air. To leave holes in the head was to let sound into the head. The mouth might have done for this, but the mouth had enough to do as it was, and, moreover, it must often be shut. In the old days, certainly, sound was conveyed to fishes in a dull way without any definite opening. But animals which live in water do not seem to use hearing much, and the sound-waves in fishes are simply conveyed through the walls of the head to the internal ear without any definite mechanism. But as soon as land-life began, owing to the changed medium through which sound-waves must now be propagated, and the new uses for sound itself, a more delicate instrument was required. And hence one of the first things attended to as the evolution went on was the construction and improvement of the ear. And this seems to have been

mainly effected by a series of remarkable developments of one of the now superfluous gill-slits.

It has long been a growing certainty to Comparative
Anatomy that the external and middle ear in Man are
simply a development, an improved edition, of the
first gill-cleft and its surrounding parts. The tym-
pano-Eustachian passage is the homologue or counterpart of the spiracle associated in the shark with the
first gill-opening. Prof. His of Leipsic has worked
out the whole development in minute detail, and con-
clusively demonstrated the mode of origin of the
external ear from the coalescence of six rounded
tubercles surrounding the first branchial cleft at an
early period of embryonic life.[1]

---

[1] Haeckel has given an earlier account of the process in the
following words :—" All the essential parts of the middle ear—the
tympanic membrane, tympanic cavity, and Eustachian tube—
develop from the first gill-opening with its surrounding parts,
which in the Primitive Fishes (Selachii) remains throughout life
as an open blow-hole, situated between the first and second gill-
arches. In the embryos of higher Vertebrates it closes in the
centre, the point of concrescence forming the tympanic membrane. The remaining outer part of the first gill-opening is the
rudiment of the outer ear-canal. From the inner part originates
the tympanic cavity, and further inward, the Eustachian tube.
In connection with these, the three bonelets of the ear develop
from the first two gill-arches ; the hammer and anvil from the
first, and the stirrup from the upper end of the second gill-arch.
Finally, as regards the external ear, the ear-shell (concha auris),
and the outer ear canal, leading from the shell to the tympanic
membrane—these parts develop in the simplest way from the skin
covering which borders the outer orifice of the first gill-opening.
At this point the ear-shell rises in the form of a circular fold of
skin, in which cartilage and muscles afterwards form."—Haeckel,
Evolution of Man, Vol. ii., p. 269.

Now, bearing in mind this theory of the origin of ears, an extraordinary corroboration confronts us. Ears are actually sometimes found bursting out *in human beings* half-way down the neck, in the exact position—namely, along the line of the anterior border of the sterno-mastoid muscle—which the gill-slits would occupy if they still persisted. In some human families, where the tendency to retain these special structures is strong, one member sometimes illustrates the abnormality by possessing the clefts alone, another has a cervical ear, while a third has both a cleft and a neck-ear—all these, of course, in addition to the ordinary ears. This cervical auricle has all the characters of the ordinary ear, " it contains yellow elastic cartilage, is skin-covered, and has muscle-fibre attached to it." [1] Dr. Sutton calls attention to the fact that on ancient statues of fauns and satyrs cervical auricles are sometimes found, and he figures the head of a satyr from the British Museum, carved long before the days of anatomy, where a sessile ear on the neck is quite distinct. A still better illustration may be seen in the Art Museum at Boston on a full-sized cast of a faun, belonging to the later Greek period ; and there are other examples in the same building. One interest of these neck-ears in statues is that they are not, as a rule, modelled after the human ear, but taken from the cervical ear of the goat, from which the general idea of the faun was derived. This shows that neck-ears were common on the goats of that period—as they are on goats to this day. The occur-

[1] Sutton, *Evolution and Disease*, p. 87.

rence of neck-ears in goats is no more than one would expect. Indeed, one would look for them not only in goats and in Man, but in all the Mammalia, for so far as their bodies are concerned all the higher animals are near relations. Observations on vestigial structures in animals are sadly wanting ; but these cervical ears are also certainly found in the horse, pig, sheep, and others.

That the human ear was not always the squat and degenerate instrument it is at present may be seen by a critical glance at its structure. Mr. Darwin records how a celebrated sculptor called his attention to a little peculiarity in the external ear, which he had often noticed both in men and women. " The peculiarity consists in a little blunt point, projecting from the inwardly folded margin or helix. When present, it is developed at birth, and, according to Professor Ludwig Meyer, more frequently in man than in woman. The helix obviously consists of the extreme margin of the ear folded inwards ; and the folding appears to be in some manner connected with the whole external ear being permanently pressed backwards. In many monkeys who do not stand high in the order, as baboons and some species of macacus, the upper portion of the ear is slightly pointed, and the margin is not at all folded inwards ; but if the margin were to be thus folded, a slight point would necessarily project towards the centre." [1] Here, then, in this discovery of the lost tip of the ancestral ear, is further and visible advertisement of Man's Descent, a surviving symbol of the stirring times and dangerous days of his animal

[1] *Descent of Man*, p. 15.

youth. It is difficult to imagine any other theory than that of Descent which could account for all these facts. That Evolution should leave such clues lying about is at least an instance of its candor.

But this does not exhaust the betrayals of this most confiding organ. If we turn from the outward ear to the muscular apparatus for working it, fresh traces of its animal career are brought to light. The erection of the ear, in order to catch sound better, is a power possessed by almost all mammals, and the attached muscles are large and greatly developed in all but domesticated forms. This same apparatus, though he makes no use of it whatever, is still attached to the ears of Man. It is so long since he relied on the warnings of hearing, that by a well-known law, the muscles have fallen into disuse and atrophied. In many cases, however, the power of twitching the ear is not wholly lost, and every school-boy can point to some one in his class who retains the capacity, and is apt to revive it in irrelevant circumstances.

One might run over all the other organs of the human body and show their affinities with animal structures and an animal past. The twitching of the ear, for instance, suggests another obsolete, or obsolescent power—the power, or rather the set of powers, for twitching the skin, especially the skin of the scalp and forehead by which we raise the eyebrows. Subcutaneous muscles for shaking off flies from the skin, or for erecting the hair of the scalp, are common among quadrupeds, and these are represented in the human subject by the still functioning muscles of the forehead, and occasionally of the head itself. Every one has met persons who possess the power of moving

the whole scalp to and fro, and the muscular apparatus
for effecting it is identical with what is normally found
in some of the Quadrumana.

Another typical vestigial structure is the *plica
semi-lunaris*, the remnant of the nictitating membrane characteristic of nearly the whole vertebrate
sub-kingdom. This membrane is a semi-transparent
curtain which can be drawn rapidly across the external surface of the eye for the purpose of sweeping
it clean. In birds it is extremely common, but it also
exists in fish, mammals, and all the other vertebrates.
Where it is not found of any functional value it is
almost always represented by vestiges of some kind.
In Man all that is left of it is a little piece of the
curtain draped at the side of the eye.

Passing from the head to the other extremity of the
body one comes upon a somewhat unexpected but
very pronounced characteristic—the relic of the tail,
and not only of the tail, but of muscles for wagging it.
Every one who first sees a human skeleton is amazed
at this discovery. At the end of the vertebral column,
curling faintly outward in suggestive fashion, are
three, four, and occasionally five vertebræ forming the
coccyx, a true rudimentary tail. In the adult this is
always concealed beneath the skin, but in the embryo,
both in Man and ape, at an early stage it is much
longer than the limbs. What is decisive as to its true
nature, however, is that even in the embryo of Man
the muscles for wagging it are still found. In the
grown-up human being these muscles are represented
by bands of fibrous tissue, but cases are known where
the actual muscles persist through life. That a distinct external tail should not still be found in Man

may seem disappointing to the evolutionist. But the want of a tail argues more for the theory of Evolution than its presence would have done. For all the anthropoids most allied to Man have long since also parted with theirs.

With regard to the presence of Hair on the body, and its disposition and direction, some curious facts may be noticed. No one, until Evolution supplied the impulse to a fresh study of the commonplace, thought it worth while to study such trifles as the presence of hair on the fingers and hands, and the slope of the hair on the arms. But now that attention is called to it, every detail is seen to be full of meaning. In all men the rudimentary hair on the arm, from the wrist to the elbow, points one way, from the elbow to the shoulder it points the opposite way. In the first case it points upwards from the wrist towards the elbow, in the other downwards from the shoulder to the elbow. This occurs nowhere else in the animal kingdom, except among the anthropoid apes and a few American monkeys, and has to do with the arboreal habit. · As Mr. Romanes, who has pointed this out, explains it, "When sitting on trees, the Orang, as observed by Wallace, places its hands above its head with its elbows pointing downwards; the disposition of hair on the arms and fore-arms then has the effect of thatch in turning the rain. Again, I find that in all species of apes, monkeys, and baboons which I have examined (and they have been numerous), the hair on the back of the hands and feet is continued as far as the first row of phalanges ; but becomes scanty, or disappears altogether, on the second row. I also find that the same peculiarity occurs in man. We

have all rudimentary hair on the first row of phalanges, both of hands and feet ; when present at all, it
is more scanty on the second row: and in no case
have I been able to find any on the terminal row.   In
all cases those peculiarities are congenital, and the
total absence or partial presence of hair on the second
phalanges is constant in different species of Quadrumana. . . .   The downward direction of the hair on
the backs of the hands is exactly the same in man as
it is in all the anthropoid apes.   Again, with regard
to hair, Darwin notices that occasionally there appear
in man a few hairs in the eyebrows much longer than
the others ; and that they seem to be a representation
of similarly long and scattered hairs which occur in
the chimpanzee, macacus, and baboon.   Lastly, about
the sixth month the human fœtus is often thickly
covered with somewhat long dark hair over the entire
body, except the soles of the feet and palms of the
hand, which are likewise bare in all quadrumanous
animals.   This covering, which is called the lanugo,
and sometimes extends even to the whole forehead,
ears, and face, is shed before birth.   So that it
appears to be useless for any purpose other than that
of emphatically declaring man a child of the
monkey."[1]   The *uselessness* of these relics, apart from
the remarkable and detailed nature of the homolo-
gies just brought out, is a circumstance very hard
to get over on any other hypothesis than that of
Descent.

Caution, of course, is required in deciding as to the
inutility of any character since its seeming uselessness

[1] *Darwin and After Darwin*, pp. 89-92.

may only mean that we do not know its use. But there are undoubtedly cases where we know that certain vestigial structures are not only useless to Man but worse than useless. Coming under this category is perhaps the most striking of all the vestigial organs, that of the Vermiform Appendix of the Cæcum. Here is a structure which is not only of no use to man now, but is a veritable death-trap. In herbivorous animals this " blind-tube " is very large—longer in some cases than the body itself—and of great use in digestion, but in Man it is shrunken into the merest rudiment, while in the Orang-outang it is only a little larger. In the human subject, owing to its diminutive size, it can be of no use whatever, while it forms an easy receptacle for the lodgment of foreign bodies, such as fruit-stones, which set up inflammation, and in various ways cause death. In Man this tube is the same in structure as the rest of the intestine; it is " covered with peritoneum, possesses a muscular coat, and is lined with mucous membrane. In the early embryo it is equal in calibre to the rest of the bowel, but at a certain date it ceases to grow *pari passu* with it, and at the time of birth appears as a thin tubular appendix to the cæcum. In the newly-born child it is often absolutely as long as in the full-grown man. This precocity is always an indication that the part was of great importance to the ancestors of the human species." [1]

So important is the key of Evolution to the modern pathologist that in cases of *malformation* his first resort is always to seek an explanation in earlier

<hr />

[1] Sutton, *Evolution and Disease*, p. 65.

forms of life. It is found that conditions which are pathological in one animal are natural in others of a lower species. When any eccentricity appears in a human body the anatomist no longer sets it down as a freak of Nature. He proceeds to match it lower down. Mr. Darwin mentions a case of a man who, in his foot alone, had no less than seven abnormal muscles. Each of these was found among the muscles of lower animals. Take, again, a common case of malformation—club-foot. All children before birth display the most ordinary form of this deformity—that, namely, where the sole is turned inwards and upwards and the foot is raised—and it is only gradually that the foot attains the normal adult position. The abnormal position, abnormal that is in adult Man, is the normal condition of things in the case of the gorilla. Club-foot, hence, is simply gorilla-foot—a case of the arrested development of a character which apparently came along the line of the direct Simian stock. So simple is this method of interpreting the present by the past, and so fruitful, that the anatomist has been able in many instances to assume the rôle of prophet. Adult man possesses no more than twelve pair of ribs; the prediction was hazarded by an older Comparative Anatomy that in the embryonic state he would be found with thirteen or fourteen. This prophecy has since been verified. It was also predicted that at this early stage he would be found to possess the insignificant remnant of a very small bone in the wrist, the so-called *os centrale*, which must have existed in the adult condition of his extremely remote ancestors. This prediction has also been fulfilled, as Weismann aptly remarks, " just as the planet Neptune was discovered after its existence had been predicted from the disturbances induced in the orbit of Uranus."[1]

But the enumeration becomes tedious. Though we are only at the beginning of the list, sufficient has been said to mark the interest of this part of the subject, and the redundancy of the proof. In the human body alone, there are at least seventy of these vestigial structures. Take away the theory that Man has evolved from a lower animal condition, and there is no explanation whatever of any one of these phenomena. With such facts before us, it is mocking human intelligence to assure us that Man has not some connection with the rest of the animal creation, or that the processes of his development stand unrelated to the other ways of Nature. That Providence, in making a new being, should deliberately have inserted these eccentricities, without their having any real connection with the things they so well imitate, or any working relation to the rest of his body is, with our present knowledge, simple irreverence.

Were it the present object to complete a proof of the descent of Man, one might go on to select from other departments of science, evidence not less striking than that from vestigial structures. From the side of palæontology it might be shown that Man appears in the earth's crust like any other fossil, and in the exact place where science would expect to find him. When born, he is ushered into life like any other animal; he is subject to the same diseases; he yields to the same treatment. When fully grown there is almost nothing in his anatomy to distinguish

[1] Weismann, *Biological Memoirs*, p. 255.

7

him from his nearest allies among other animals—
almost bone for bone, nerve for nerve, muscle for
muscle he is the same.   There is in fact a body of
evidence now before science for the animal origin of
Man's physical frame which it is impossible for a
thinking mind to resist.   Up to this point two only
out of the many conspiring lines of testimony have
been drawn upon for their contribution ; but enough
has been said to encourage us, with this as at least a
working theory, to continue the journey.   It is the
Ascent of Man that concerns us and not the Descent.
And these amazing facts about the past are cited for
a larger purpose than to produce conviction on a point
which, after all, is of importance only in its higher
implications.

CHAPTER III.

THE ARREST OF THE BODY.

" On the Earth there will never be a higher Creature than Man." [1] It is a daring prophecy, but every probability of Science attests the likelihood of its fulfilment. The goal looked forward to from the beginning of time has been attained. Nature has succeeded in making a Man; she can go no further; Organic Evolution has done its work.

This is not a conceit of Science, nor a reminiscence of the pre-Copernican idea that the centre of the universe is the world, and the centre of the world Man. It is the sober scientific probability that with the body of Man the final fruit of the tree of Organic Evolution has appeared ; that the highest possibilities open to flesh and bone and nerve and muscle have now been realized ; that in whatever direction, and with whatever materials, Evolution still may work, it will never produce any material thing more perfect in design or workmanship ; that in Man, in short, about this time in history, we are confronted with a stupendous crisis in Nature,—the Arrest of the Animal.

[1] Fiske, *Destiny of Man*, p. 26. What follows owes much to this suggestive *brochure*.

The Man, the Animal Man, the Man of Organic Evolution, it is at least certain, will not go on. It is another Man who will go on, a Man within this Man; and that he may go on the first Man must stop. Let us try for a moment to learn what it is to stop. Nothing could teach Man better what is meant by his going on.

One of the most perfect pieces of mechanism in the human body is the Hand. How long it has taken to develop may be dimly seen by a glance at the long array of less accurate instruments of prehension which shade away with ever decreasing delicacy and perfectness as we descend the scale of animal life. At the bottom of that scale is the Amœba. It is a speck of protoplasmic jelly, headless, footless, and armless. When it wishes to seize the microscopic particle of food on which it lives a portion of its body lengthens out, and, moving towards the object, flows over it, engulfs it, and melts back again into the body. This is its Hand. At any place, and at any moment, it creates a Hand. Each Hand is extemporized as it is needed; when not needed it is not. Pass a little higher up the scale and observe the Sea-Anemone. The Hand is no longer extemporized as occasion requires, but lengthened portions of the body are set apart and kept permanently in shape for the purpose of seizing food. Here, in the capital of twining tentacles which crowns the quivering pillar of the body, we get the rude approximation to the most useful portion of the human Hand—the separated fingers. It is a vast improvement on the earlier Hand, but the jointless digits are still imperfect; it is simply the Amœba Hand cut into permanent strips.

Passing over a multitude of intermediate forms, watch, in the next place, the Hand of an African Monkey. Note the great increase in usefulness due to the muscular arm upon which the Hand is now extended, and the extraordinary capacity for varied motion afforded by the threefold system of jointing at shoulder, elbow, and wrist. The Hand itself is almost the human Hand; there are palm and nail and articulated fingers. But observe how one circumstance hinders the possessor from taking full advantage of these great improvements,—this Hand has no thumb, or if it has, it is but a rudiment. To estimate the importance of this apparently insignificant organ, try for a moment without using the thumb to hold a book, or write a letter, or do any single piece of manual work. A thumb is not merely an additional finger, but a finger so arranged as to be *opposable to the other fingers*, and thus possesses a practical efficacy greater than all the fingers put together. It is this which gives the organ the power to seize, to hold, to manipulate, to do higher work; this simple mechanical device in short endows the Hand of intelligence with all its capacity and skill. Now there are animals, like the Colobi, which have no thumb at all; there are others, like the Marmoset, which possess the thumb, but in which it is not opposable; and there are others, the Chimpanzee for instance, in which the Hand is in all essentials identical with Man's. In the human form the thumb is a little longer, and the whole member more delicate and shapely, but even for the use of her highest product, Nature has not been able to make anything much more perfect than the hand of this anthropoid ape.

Is the Hand then finished? Can Nature take out no new patent in this direction? Is the fact that no novelty is introduced in the case of Man a proof that the ultimate Hand has appeared? By no means. And yet it is probable for other reasons that the ultimate Hand has appeared; that there will never be a more perfectly handed animal than Man. And why? Because the causes which up to this point have furthered the evolution of the Hand have begun to cease to act. In the perfecting of the bodily organs, as of all other mechanical devices, necessity is the mother of invention. As the Hand was given more and more to do, it became more and more adapted to its work. Up to a point, it responded directly to each new duty that was laid upon it. But only up to a point. There came a time when the necessities became too numerous and too varied for adaptation to keep pace with them. And the fatal day came, the fatal day for the Hand, when he who bore it made a new discovery. It was the discovery of Tools. Henceforth what the Hand used to do, and was slowly becoming adapted to do better, was to be done by external appliances. So that if anything new arose to be done, or to be better done, it was not a better Hand that was now made but a better tool. Tools are external Hands. Levers are the extensions of the bones of the arm. Hammers are callous substitutes for the fist. Knives do the work of nails. The vice and the pincers replace the fingers. The day that Cave-man first split the marrow bone of a bear by thrusting a stick into it, and striking it home with a stone—that day the doom of the Hand was sealed.

But has not Man to make his tools, and will not that induce the development of the Hand to an as yet unknown perfection? No. Because tools are not made with the Hand. They are made with the Brain. For a time, certainly, Man had to make his tools, and for a time this work recompensed him physically, and the arm became elastic and the fingers dexterous and strong. But soon he made tools to make these tools. In place of shaping things with the Hand, he invented the turning-lathe; to save his fingers he requisitioned the loom; instead of working his muscles he gave out the contract to electricity and steam. Man, therefore, from this time forward will cease to develop materially these organs of his body. If he develops them outside his body, filling the world everywhere with artificial Hands, supplying the workshops with fingers more intricate and deft than Organic Evolution could make in a millennium, and loosing energies upon them infinitely more gigantic than his muscles could generate in a lifetime, it is enough. Evolution after all is a slow process. Its great labor is to work up to a point where Invention shall be possible, and where, by the powers of the human mind, and by the mechanical utilization of the energies of the universe, the results of ages of development may be anticipated. Further changes, therefore, within the body itself are made unnecessary. Evolution has taken a new departure. For the Arrest of the Hand is not the cessation of Evolution but its immense acceleration, and the re-direction of its energies into higher channels.

Take up the functions of the animal body one by one, and it will be seen how the same arresting finger is laid upon them all. To select an additional illus-

tration, consider the power of Sight. Without pausing to trace the steps by which the Eye has reached
its marvellous perfection, or to estimate the ages spent
in polishing its lenses and adjusting the diaphragms
and screws, ask the simple question whether, under
the conditions of modern civilization, anything now is
being added to its quickening efficiency, or range. Is
it not rather the testimony of experience that if anything its power has begun to wane? Europe even
now affords the spectacle of at least one nation so
short-sighted that it might almost be called a myopic
race. The same causes, in fact, that led to the Arrest
of the Hand are steadily working to stop the development of the Eye. Man, when he sees with difficulty,
does not now improve his Eye; he puts on a *pince-nez.*
Spectacles—external eyes—have superseded the work
of Evolution. When his sight is perfect up to a point,
and he desires to examine objects so minute as to lie
beyond the limit of that point, he will not wait for
Evolution to catch up upon his demand and supply
him, or his children's children, with a more perfect
instrument. He will invest in a microscope. Or
when he wishes to extend his gaze to the moon and
stars, he does not hope to reach to-morrow the distances which to-day transcend him. He invents the
telescope. Organic Evolution has not even a chance.
In every direction the external eye has replaced the
internal, and it is even difficult to suggest where any
further development of this part of the animal can
now come in. There are still, and in spite of all
instruments, regions in which the unaided organs of
Man may continue to find a field for the fullest exercise, but the area is slowly narrowing, and in every

direction the appliances of Science tempt the body to
accept those supplements of the Arts, which, being
accepted, involve the discontinuance of development
for all the parts concerned. Even where a mechanical
appliance, while adding range to a bodily sense, has
seemed to open a door for further improvement, some
correlated discovery in a distant field of science, as by
some remorseless fate, has suddenly taken away the
opportunity and offered to the body only an additional
inducement for neglect. Thus it might be thought
that the continuous use of the telescope, in the attempt to discover more and more indistinct and distant heavenly bodies, might tend to increase the efficiency of the Eye. But that expectation has vanished
already before a further fruit of Man's inventive
power. By an automatic photographic apparatus
fixed to the telescope, an Eye is now created vastly
more delicate and in many respects more efficient than
the keenest eye of Man. In at least five important
particulars the Photographic Eye is the superior of
the Eye of Organic Evolution. It can see where the
human Eye, even with the best aids of optical instruments, sees nothing at all; it can distinguish certain
objects with far greater clearness and definition;
owing to the rapidity of its action it can instantly detect changes which are too sudden for the human eye
to follow; it can look steadily for hours without growing tired; and it can record what it sees with infal-
lible accuracy upon a plate which time will not efface.
How long would it take Organic Evolution to arrive
at an Eye of such amazing quality and power? And
with such a piece of mechanism available, who, rather
than employ it even to the neglect of his organs of

vision, would be content to await the possible attain-
ment of an equal perfection by his descendants some
million years hence? Is there not here a conspicuous
testimony to the improbability of a further Evolution
of the sense of Sight in civilized communities—in
other words, another proof of the Arrest of the
Animal? What defiance of Evolution, indeed, what
affront to Nature, is this? Man prepares a complicated telescope to supplement the Eye created by Evolution, and no sooner is it perfected than it occurs to
him to create another instrument to aid the Eye in
what little work is left for it to do. That is to say,
he first makes a mechanical supplement to his Eye,
then constructs a mechanical Eye, which is better
than his own, to see through it, and ends by discard-
ing, for many purposes, the Eye of Organic Evolution
altogether.

As regards the other functions of civilized Man,
the animal in almost every direction has reached
its maximum. Civilization—and the civilized state, be
it remembered, is the ultimate goal of every race and
nation—is always attended by deterioration of some
of the senses. Every man pays a definite price or
forfeit for his taming. The sense of smell, compared
with its development among the lower animals, is in
civilized Man already all but gone. Compared even
with a savage, it is an ascertained fact that the civilized Man in this respect is vastly inferior. So far as
hearing is concerned, the main stimulus—fear of surprise by enemies—has ceased to operate, and the
muscles for the erection of the ears have fallen into
disuse. The ear itself in contrast with that of the
savage is slow and dull, while compared with the

quick sense of the lower animals, the organ is almost deaf. The skin, from the continuous use of clothes, has forfeited its protective power. Owing to the use of viands cooked, the muscles of the jaw are rapidly losing strength. The teeth, partly for a similar reason, are undergoing marked degeneration. The third molar, for instance, among some nations is already showing symptoms of suppression, and that this threatens ultimate extinction may be reasoned from the fact that the anthropoid apes have fewer teeth than the lower monkeys, and these fewer than the preceding generation of insectivorous mammals.

In an age of vehicles and locomotives the lower limbs find their occupation almost gone. For mere muscle, that on which his whole life once depended, Man has almost now no use. Agility, nimbleness, strength, once a stern necessity, are either a luxury or a pastime. Their outlet is the cricket-field or the tennis-court. To keep them up at all artificial means —dumb-bells, parallel-bars, clubs—have actually to be devised. Vigor of limb is not to be found in common life, we look for it in the Gymnasium; agility is relegated to the Hippodrome. Once all men were athletes; now you have to pay to see them. More or less with all the animal powers it is the same. To some extent at least some phonograph may yet speak for us, some telephone hear for us, the typewriter write for us, chemistry digest for us, and incubation nurture us. So everywhere the Man as Animal is in danger of losing ground. He has expanded until the world is his body. The former body, the hundred and fifty pounds or so of organized tissue he carries about with him, is little more than a mark of identity.

It is not *he* who is there, he cannot be there, or anywhere, for he is everywhere. The material part of
him is reduced to a symbol; it is but a link with the
wider framework of the Arts, a belt between machinery and machinery. His body no longer gener-
ates, but only utilizes energy; alone he is but a tool,
a medium, a turncock of the physical forces.

Now with what feelings do we regard all this? Is
not the crowning proof of the thesis under review that
we watch this evidence accumulating against the body
with no emotion and hear the doom of our clay
pronounced without a regret? It is nothing to aspir-
ing Man to watch the lower animals still perfecting
their mechanism and putting all his physical powers
and senses to the shame. It is nothing to him to be
distanced in nimbleness by the deer: has he not his
bullet? Or in strength by the horse: has he not bit
and bridle? Or in vision by the eagle: his field-glass
out-sees it. How easily we talk of the body as a
thing without us, as an impersonal *it*. And how naturally when all is over, do we advertise its irrelevancy
to *ourselves* by consigning its borrowed atoms to the
anonymous dust. The fact is, in one aspect, the body,
to Intelligence, is all but an absurdity. One is almost
ashamed to have one. The idea of having to feed it,
and exercise it, and humor it, and put it away in the
dark to sleep, to carry it about with one everywhere,
and not only it but its wardrobe—other material
things to make this material thing warm or keep it
cool—the whole situation is a comedy. But judge
what it would be if this exacting organism went on
evolving, multiplied its members, added to its intricacy, waxed instead of waned? So complicated is

it already that one shrinks from contemplating a future race having to keep in repair an apparatus more involved and delicate. The practical advantage is enormous of having all improvements henceforth external, of having insensate organs made of iron and steel rather than of wasting muscle and palpitating nerve. For these can be kept at no physiological cost, they cannot impede the other machinery, and when that finally comes to the last break-down there will be the fewer wheels to stop.

So great indeed is the advantage of increasing mechanical supplements to the physical frame rather than exercising the physical frame itself, that this will become nothing short of a temptation; and not the least anxious task of future civilization will be to prevent degeneration beyond a legitimate point, and keep up the body to its highest working level. For the first thing to be learned from these facts is not that the Body is nothing and must now decay, but that it is most of all and more than ever worthy to be preserved. The moment our care of it slackens, the Body asserts itself. It comes out from under arrest—which is the one thing to be avoided. Its true place by the ordained appointment of Nature is where it can be ignored; if through disease, neglect or injury it returns to consciousness, the effect of Evolution is undone. Sickness is degeneration; pain the signal to resume the evolution. On the one hand, one must "reckon the Body dead"; on the other, one must think of it in order not to think of it.

This arrest of physical development at a specific point is not confined to Man. Everywhere in the organic world science is confronted with arrested types. While endless groups of plant and animal forms have advanced during the geological ages, other whole groups have apparently stood still—stood still, that is to say, not in time but in organization. If Nature is full of moving things, it is also full of fixtures. Thirty-one years ago Mr. Huxley devoted the anniversary Address of the Geological Society to a consideration of what he called "Persistent Types of Life," and threw down to Evolutionists a puzzle which has never yet been fully solved. While some forms attained their climacteric tens of thousands of years ago and perished, others persevered, and, without advancing in any material respect, are alive to this day. Among the most ancient Carboniferous plants, for instance, are found certain forms generically identical with those now living. The cone of the existing Araucaria is scarcely to be distinguished from that of an Oolite form. The Tabulate Corals of the Silurian period are similar to those which exist to-day. The Lamp-shells of our present seas so abounded at the same ancient date as to give their name to one of the great groups of Silurian rocks—the Lingula Flags. Star-fishes and Sea-urchins, almost the same as those which tenant the coast-lines of our present seas, crawled along what are now among the most ancient fossiliferous rocks. Both of the forms just named, the Brachiopods and the Echinoderms, have come down to us almost unchanged through the nameless gap of time which separates the Silurian and Old Red Sandstone periods from the present era.

This constancy of structure reveals a conservatism in Nature, as unexpected as it is wide-spread. Does it mean that the architecture of living things has a limit

beyond which development cannot go ? Does it mean
that the morphological possibilities along certain lines
of bodily structure have exhausted themselves, that
the course of conceivable development in these instances has actually run out? In Gothic Architecture, or in Norman, there are terminal points which,
once reached, can be but little improved upon. Without limiting working efficiency, they can go no further.
These styles in the very nature of things seem to have
limits. Mr. Ruskin has indeed assured us that there
are only three possible forms of good architecture in
the world; Greek, the architecture of the Lintel;
Romanesque, the architecture of the Rounded Arch;
Gothic, the architecture of the Gable. "All the architects in the world will never discover any other way
of bridging a space than these three, the Lintel, the
Round Arch, the Gable; they may vary the curve of
the arch, or curve the sides of the gable, or break
them down; but in doing this they are merely modifying or sub-dividing, not adding to the generic
form." [1]

In some such way, there may be terminal generic
forms in the architecture of animals; and the persistent types just named may represent in their several
directions the natural limits of possible modification.
No further modification of a radical kind, that is to
say, could in these instances be introduced without detriment to practical efficiency. These terminal forms thus mark a normal maturity, a goal;
they represent the ends of the twigs of the tree of
life.

Now consider the significance of that fact. Nature

[1] *Stones of Venice*, II. 236.

is not an interminable succession. It is not always a becoming. Sometimes things arrive. The Lamp-shells have arrived, they are part of the permanent furniture of the world; along that particular line, there will probably never be anything higher. The Star-fishes also have arrived, and the Sea-urchins, and the Nautilus, and the Bony Fishes, the Tapirs, and possibly the Horse—all these are highly divergent forms which have run out the length of their tether and can go no further. When the plan of the world was made, to speak teleologically, these types of life were assigned their place and limit, and there they have remained. If it were wanted to convey the impression that Nature had some large end in view, that she was not drifting aimlessly towards a general higher level, it could not have been done more impressively than by everywhere placing on the field of Science these fixed points, these innumerable consummations, these clean-cut mountain peaks, which for millenniums have never grown. Even as there is a plan in the parts, there is a plan in the whole.

But the most certain of all these "terminal points" in the evolution of Creation is the body of Man. Anatomy places Man at the head of all other animals that were ever made; but what is infinitely more instructive, with him, as we have just seen, the series comes to an end. Man is not only the highest branch, but the highest possible branch. Take as a last witness the testimony of anatomy itself with regard to the human brain. Here the fact is not only reaffirmed but the rationale of it suggested in terms of scientific law. "The development of the brain is in connection with a whole system of development of the

head and face which cannot be carried further than
in Man. For the mode in which the cranial cavity is
gradually increased in size is a regular one, which
may be explained thus: we may look on the skull as
an irregular cylinder, and at the same time that it is
expanded by increase of height and width it also
undergoes a curvature or bending on itself, so that the
base is crumpled together while the roof is elongated.
This curving has gone on in Man till the fore end of
the cylinder, the part on which the brain rests above
the nose, is nearly parallel to the aperture of communication of the skull with the spinal canal, *i. e.,* the
cranium has a curve of 180° or a few degrees more or
less. This curving of the base of the skull involves
change in position of the face bones also, and could
not go on to a further extent without cutting off the
nasal cavity from the throat . . . Thus there is
anatomical evidence that the development of the vertebrate form has reached its limit by completion in
Man." [1]

This author's conception of the whole field of living
nature is so suggestive that we may continue the quo-
tation: "To me the animal kingdom appears not in
indefinite growth like a tree, but a temple with many
minarets, none of them capable of being prolonged—
while the central dome is completed by the structure
of man. The development of the animal kingdom is
the development of intelligence chained to matter;
the animals in which the nervous system has reached
the greatest perfection are the vertebrates, and in Man
that part of the nervous system which is the organ of

[1] Prof. J. Cleland, M.D., F.R.S., *Journal of Anatomy,* Vol.
xviii., pp. 360-1.

intelligence reaches, as I have sought to show, the
highest development possible to a vertebrate animal,
while intelligence has grown to reflection and volition.
On these grounds, I believe, not that Man is the
highest possible intelligence, but that the human body
is the highest form of human life possible, subject to
the conditions of matter on the surface of the globe,
and that the structure completes the design of the animal kingdom." [1]

Never was the body of Man greater than with this
sentence of suspension passed on it, and never was
Evolution more wonderful or more beneficent than
when the signal was given to stop working at Man's
animal frame. This was an era in the world's history.
For it betokened nothing less than that the cycle of
matter was now complete, and the one prefatory task
of the ages finished. Henceforth the *Weltanschauung*
is forever changed. From this pinnacle of matter is
seen at last what matter is for, and all the lower lives
that ever lived appear as but the scaffolding for this
final work. The whole sub-human universe finds its
reason for existence in its last creation, its final justifi-
cation in the new immaterial order which opened with
its close. Cut off Man from Nature, and, metaphysical necessity apart, there remains in Nature no
divinity. To include Man in Evolution is not to lower
Man to the level of Nature, but to raise Nature to his
high estate. There he was made, these atoms are his
confederates, these plant cells raised him from the
dust, these travailing animals furthered his Ascent:
shall he excommunicate them now that their work is
done? Plant and animal have each their end, but

[1] *Journal of Anatomy*, Vol. XVIII., p. 362.

Man is the end of all the ends. The latest science re-instates him, where poet and philosopher had already placed him, as at once the crown, the master, and the rationale of creation. "Not merely," says Kant, "is he like all organized beings an end in nature, but also here on earth the last end of nature, in reference to whom all other natural things constitute a system of ends." Yet it is not because he is the end of ends, but the beginning of beginnings, that the completion of the Body marks a crisis in the past. At last Evolution had culminated in a creation so complex and exalted as to form the foundation for an inconceivably loftier super-organic order. The moment an organism was reached through which Thought was possible, nothing more was required of matter. The Body was high enough. Organic Evolution might now even resign its sovereignty of the world; it had made a thing which was now its master. Henceforth Man should take charge of Evolution even as up till now he had been the one charge of it. Henceforth his selection should replace Natural Selection; his judgment guide the struggle for life; his will determine for every plant upon the earth, whether it should bloom or fade, for every animal whether it should increase, or change, or die. So Man entered into his Kingdom.

Science is charged, be it once more recalled, with numbering Man among the beasts, and levelling his body with the dust. But he who reads for himself the history of creation as it is written by the hand of Evolution will be overwhelmed by the glory and honor heaped upon this creature. To be a Man, and to have no conceivable successor; to be the fruit and

crown of the long past eternity, and the highest possible fruit and crown; to be the last victor among the
decimated phalanxes of earlier existences, and to be
nevermore defeated; to be the best that Nature in her
strength and opulence can produce; to be the first of
that new order of beings who by their dominion over
the lower world and their equipment for a higher,
reveal that they are made in the Image of God—to be
this is to be elevated to a rank in Nature more exalted
than any philosophy or any poetry or any theology
have ever given to Man.   Man was always told that
his place was high; the reason for it he never knew
till now; he never knew that his title deeds were the
very laws of Nature, that he alone was the Alpha and
Omega of Creation, the beginning and the end of
Matter, the final goal of Life.

Nature is full of new departures; but never since
time began was there anything approaching in importance that period when the slumbering animal, Brain,
broke into intelligence, and the Creature first felt that
it had a Mind.   From that dateless moment a higher
and swifter progress of the world began.   Henceforth,
Intelligence triumphed over structural adaptation.
The wise were naturally selected before the strong.
The Mind discovered better methods, safer measures,
shorter cuts.   So the body learned to refer to it, then
to defer to it.   As the Mind was given more to do, it
enlarged and did its work more perfectly.   Gradually
the favors of Evolution—exercise, alteration, differentiation, addition—which were formerly distributed promiscuously among the bodily organs—were
now lavished mainly upon the Brain.   The gains
accumulated with accelerating velocity; and by sheer

superiority and fitness for its work, the Intellect rose
to commanding power, and entered into final possession of a monopoly which can never be disturbed.

Now this means not only that an order of higher
animals has appeared upon the earth, but that an
altogether new page in the history of the universe has
begun to be written. It means nothing less than that
the working of Evolution has changed its course.
Once it was a physical universe, now it is a psychical
universe. And to say that the working of Evolution
has changed its course, and set its compass in psychical directions, is to call attention to the most
remarkable fact in Nature. Nothing so original or so
revolutionary has ever been given to science to discover, to ponder, or to proclaim. The power of this
event to strike and rouse the mind will depend upon
one's sense of what the working of Evolution has
been to the world; but those who realize this even
dimly will see that no emphasis of language can exaggerate its significance. Let imagination do its best to
summon up the past of Nature. Beginning with the
panorama of the Nebular Hypothesis, run the eye over
the field of Palæontology, Geology, Botany, and
Zoology. Watch the majestic drama of Creation
unfolding, scene by scene and act by act. Realize
that one power, and only one, has marshalled the
figures for this mighty spectacle; that one hand, and
only one, has carried out these transformations; that
one principle, and only one, has controlled each subsidiary plot and circumstance; that the same great
patient unobtrusive law has guided and shaped the
whole from its beginnings in bewilderment and chaos
to its end in order, harmony, and beauty. Then watch

the curtain drop.   And as it moves to rise again,
behold the new actor upon the stage.   Silently, as all
great changes come, Mental Evolution has succeeded
Organic.   All the things that have been now lie in the
far background as forgotten properties.   And Man
stands alone in the foreground, and a new thing,
Spirit, strives within him.

CHAPTER IV.

THE DAWN OF MIND.

THE most beautiful witness to the Evolution of Man is the Mind of a little child. The stealing in of that inexplicable light—yet not more light than sound or touch—called consciousness, the first flicker of memory, the gradual governance of will, the silent ascendancy of reason—these are studies in Evolution the oldest, the sweetest, and the most full of meaning for mankind. Evolution, after all, is a study for the nursery. It was ages before Darwin or Lamarck or Lucretius that Maternity, bending over the hollowed cradle in the forest for a first smile of recognition from her babe, expressed the earliest trust in the doctrine of development. Every mother since then is an unconscious Evolutionist, and every little child a living witness to Ascent.

Is the Mind a new or an old thing in the world? Is it an Evolution from beneath or an original gift from heaven? Did the Mind, in short, come down the ages like the Body, and does the mother's faith in the intellectual unfolding of her babe include a remoter origin for all human faculty? Let the mother look at her child and answer. " It is the very breath of God," she says; " this Child-Life is Divine." And she is right. But let her look again. That forehead, whose is it ? It is hers. And the frown which darkened it just now? Is hers also. And that which caused the frown to darken, that something or nothing, behind the forehead, that flash of pride, or scorn, or hate? Alas, it is her very own. And as the years roll on, and the budding life unfolds, there is scarcely a mood or gesture or emotion that she does not know is borrowed. But whence in turn did she receive them? From an earlier mother. And she ? From a still earlier mother. And she ? From the savage-mother in the woods. And the savage-mother ?

Shall we hesitate here ? We well may. So Godlike a gift is intellect, so wondrous a thing is consciousness, that to link them with the animal world seems to trifle with the profoundest distinctions in the universe. Yet to associate these supersensuous things with the animal kingdom is not to identify them with the animal-body. Electricity is linked with metal rods, it is not therefore metallic. Life is associated with protoplasm, it is not therefore albuminous. Instinct is linked with matter, but it is not therefore material; Intellect with animal matter, but is not therefore animal. As we rise in the scale of Nature we encounter new orders of phenomena, Matter, Life, Mind, each higher than that before it, each totally and forever different, yet each using that beneath it as the pedestal for its further progress. Associated with animal-matter—how associated no psychology, no physiology, no materialism, no spiritualism, has even yet begun to hint—may there not have been from an early dawn the elements of a future Mind? Do the wide analogies of Nature not make the suggestion worthy at least of inquiry? The fact, to which there is no exception, that all lesser things evolve, the suggestion, which is daily growing into a further certainty, that there is a mental evolution among animals from the Cœlenterate to the Ape; the fact that the unfolding of the Child-Mind is itself a palpable evolution; the infinitely more significant circumstance that the Mind in a child seems to unfold in the order in which it would unfold if its mental faculties were received from the Animal world, and in the order in which they have already asserted themselves in the history of the race. These seem formidable facts on the side of those consistent evolutionists who, in the face of countless difficulties and countless prejudices, still press the lawful inquiry into the development of human faculty.

The first feeling in most minds when the idea of mental evolution is presented, is usually one of amusement. This not seldom changes, when the question is seen to be taken seriously, into wonder at the daring of the suggestion or pity for its folly. All great problems have been treated in this way. All have passed through the inevitable phases of laughter, contempt, opposition. It ought to be so. And if this problem is "perhaps the most interesting that has ever been submitted to the contemplation of our race," [1] its basis cannot be criticised with too great care. But none have a right to question either the sanity or the sanctity of such investigations, still less to dismiss them idly on *a priori* grounds, till they have approached the practical problem for themselves, and

[1] Romanes, *Mental Evolution in Man*, p. 2.

heard at least the first few relevant words from Nature. For one has only to move for a little among the facts to see what a world of interest lies here, and to be forced to hold the judgment in suspense till the sciences at work upon the problem have further shaped their verdict. Thinkers who are entitled to respect have even gone further. They include mental evolution not only among the hypotheses of Science but among its facts and its necessary facts. " Is it conceivable," asks Mr. Romanes, " that the human mind can have arisen by way of a natural genesis from the minds of the higher quadrumana? I maintain that the material now before us is sufficient to show, not only that this is conceivable, but inevitable." [1]

It is no part of the present purpose to discuss the ultimate origin or nature of Mind. Our subject is its development. At the present moment the ultimate origin of Mind is as inscrutable a mystery as the origin of Life. It is sometimes charged against Evolution that it tries to explain everything and to rob the world of all its problems. There does not appear the shadow of a hope that it is about to rob it of this. On the contrary the foremost scientific exponents of the theory of mental evolution are ceaselessly calling attention to the inscrutable character of the element whose history they attempt to trace. " On the side of its philosophy," says Mr. Romanes, " no one can have a deeper respect for the problem of self-consciousness than I have; for no one can be more profoundly convinced than I am that the problem on this side does not admit of solution. In other

[1] *Op. cit.*, p. 213.

words, so far as this aspect of the matter is concerned, I am in complete agreement with the most advanced idealist. I am as far as any one can be from throwing light upon the intrinsic nature of the probable origin of that which I am endeavoring to trace." [1] Mr. Darwin himself recoiled from a problem so transcendent: "I have nothing to do with the origin of the mental powers, any more than I have with that of life itself." [2] "In what manner," he elsewhere writes, "the mental powers were first developed in the lowest organisms, is as hopeless an inquiry as how life itself first originated." [3]

Notwithstanding his appreciation of the difficulty of the ultimate problem, Mr. Darwin addressed his whole strength to the question of the Evolution of Mind—the Evolution as distinguished from its origin and nature; and in this he has recently had many followers, as well as many opponents. Among the latter stand the co-discoverer with him of Natural Selection, Mr. Alfred Russel Wallace, and Mr. St. George Mivart. Mr. Wallace's opposition, from a scientific point of view, is not so hostile, however, as is generally supposed. While holding his own view as to the origin of Mind, what he attacks in Mr. Darwin's theory of mental evolution is, not the development itself, but only the supposition that it could have been due to Natural Selection. Mr. Wallace's authority is frequently quoted to show that the mathematical, the musical and the artistic faculties could not have been evolved, whereas all he has really emphasized is that "they could not have been devel-

[1] *Mental Evolution in Man*, pp. 194-5.
[2] *Origin of Species*, p. 191.   [3] *Descent of Man*, p. 66.

oped under the law of Natural Selection." [1]   In short,
the conclusion of Mr. Darwin which his colleague
found "not to be supported by adequate evidence, and
to be directly opposed to many well-ascertained facts,"
was not a general theorem, but a specific one.   And
many will agree with Mr. Wallace in doubting "that
man's entire nature and all his faculties, whether
moral, intellectual, or spiritual, have been derived
from their rudiments in the lower animals, *in the same
manner and by the action of the same general laws* as
his physical structure has been derived." [2]

The more this problem has been investigated, the
difficulties of the whole field increase, and the off-hand
acceptance of any specific evolution theory finds less
and less encouragement.   No serious thinker, on
whichever side of the controversy, has succeeded in
lessening to his own mind the infinite distance between the Mind of Man and everything else in Nature,
and even the most consistent evolutionists are as
unanimous as those who oppose them, in their assertion of the uniqueness of the higher intellectual
powers.   The consensus of scientific opinion here is
extraordinary.   " I know nothing," says Huxley, in
the name of biology, "and never hope to know
anything, of the steps by which the passage from
molecular movement to states of consciousness is
effected." [3]   " The two things," emphasizes the physi-
cist, "are on two utterly different platforms, the
physical facts go along by themselves, and the mental facts go along by themselves." [4]   " It is all through

[1] *Darwinism*, p. 469.   [2] *Ibid.*, p. 461.
[3] *Contemporary Review*, 1871.
[4] Clifford, *Fortnightly Review*, 1874.

and forever inconceivable," protests the German physiologist, " that a number of atoms of Carbon, Hydrogen, Nitrogen, Oxygen, and so on, shall be other than indifferent as to how they are disposed and how they move, how they were disposed and how they moved, how they will be disposed and how they will be moved. It is utterly inconceivable how consciousness shall arise from their joint action." [1] So impressed is even Mr. Lloyd Morgan, mental evolutionist though he be, with the gap between the Minds of Man and brute that his language is almost as strong: " I for one do not for a moment question that the mental processes of man and animals are alike products of evolution. The power of cognizing relations, reflection and introspection, appear to me to mark a new departure in evolution," [2] and " I am not prepared to say that there is a difference in kind between the mind of man and the mind of a dog. This would imply a difference in origin or a difference in the essential nature of its being. There is a great and marked difference in kind between the material processes which we call physiological and the mental processes we call psychical. They belong to wholly different orders of being. I see no reason for believing that mental processes in man differ thus in kind from mental processes in animals. But I do think that we have, in the introduction of the analytic faculty, so definite and marked a new departure that we should emphasize it by saying that the faculty of perception, in its various specific grades, differs generically from the faculty of concep-

[1] Du Bois-Reymond, *Ueber die Grenzen des Naturerkennens,* p. 42.
[2] C. Lloyd Morgan, *Nature*, Sept. 1, 1892, p. 417.

tion. And believing, as I do, that conception is beyond the power of my favorite and clever dog, I am forced to believe that his mind differs generically from my own." [1]

Should any one feel it necessary either to his view of Man or of the Universe to hold that a great gulf lies here, it is open to him to cling to his belief. The present thesis is simply that Man has ascended. After all, little depends on whether the slope is abrupt or gentle, whether Man reaches the top by a uniform flight or has here and there by invisible hands to be carried across a bridgeless space. In any event it is Nature's staircase. To say that self-consciousness has arisen from sensation, and sensation from the function of nutrition, let us say, in the *Mimosa pudica* or Sensitive Plant, may be right or wrong; but the error can only be serious when it is held that that accounts either for self-consciousness or for the transition. Mimosa can be defined in terms of Man; but Man cannot be defined in terms of Mimosa. The first is possible because there is the least fraction in that which is least in Man of that which is greatest in Mimosa; the last is impossible because there is nothing in Mimosa of that which is greatest in Man. What the two possess in common, or seem to possess, may be a basis for comparison, for what it is worth; but to include in the comparison the ninety-nine and nine-tenths per cent. of what is over and above that common fraction is by no sort of reasoning lawful. Man, in the last resort, has self-consciousness, Mimosa sensation; and the difference is qualitative as well as quantitative.

If, however, it is a fallacy to ignore the qualitative

[1] C. Lloyd Morgan, *Animal Life and Intelligence*, p. 350.

differences arising in the course of the transition, it may be a mistake, on the other hand, to make nothing of the transition. If in the name of Science the advocate of the Law of Continuity demands that it be rectified, he may well make the attempt. The partial truth for the present perhaps amounts to this, that earlier phases of life exhibit imperfect manifestations of principles which in the higher structure and widened environment of later forms are more fully manifested and expressed, yet are neither contained in the earlier phases nor explained by them. At the same time, everything that enters into Man, every sensation, emotion, volition, enters with a difference, a difference due to the fact that he is a rational and self-conscious being, a difference therefore which no emphasis of language can exaggerate. The music varies with the ear ; varies with the soul behind the ear ; relates itself with all the music that ear has ever heard before ; with the mere fact that what that ear hears, it hears as music ; that it hears at all ; that it knows that it hears. Man differs from every other product of the evolutionary process in being able to see that it is a process, in sharing and rejoicing in its unity, and in voluntarily working through the process himself. If he is part of it he is also more than part of it, since he is at once its spectator, its director, and its critic. " Even on the hypothesis of a psychic life in all matter we come to an alteration indeed, but not an abolition, of the contrast between body and soul. Of course on that hypothesis they are distinguished by no qualitative difference in their natures, but still less do they blend into one ; the one individual ruling soul always remains facing, in an attitude of complete isolation, the homogeneous but ministrant monads, the joint multitude of which forms the living body." [1]

With these preliminary cautions, let us turn for a little to the facts. The field here is so full of interest in itself that apart from its forming a possible chapter in the history of Man it is worth a casual survey.

The difficulty of establishing even the general question of Ascent is of course obvious. After Mind emerged from the animal state, for a long time, and in the very nature of the case, no record of its progress could come down to us. The material Body has left its graduated impress upon the rocks in a million fossil forms; the Spirit of Man, at the other extreme of time, has traced its ascending curve on the tablets of civilization, in the drama of history, and in the monuments of social life; but the Mind must have risen into its first prominence during a long, silent and dateless interval which preceded the era of monumental records. Mind cannot be exhumed by Palæ-ontology or fully embalmed in unwritten history, and apart from the analogies of Embryology we have nothing but inference to guide us until the time came when it was advanced enough to leave some tangible register behind.

But so far as knowledge is possible there are mainly five sources of information with regard to the past of Mind. The first is the Mind of a little child; the second the Mind of lower animals; the third, those material witnesses—flints, weapons, pottery—to primitive states of Mind which are preserved in anthropological museums; the fourth is the **Mind of a Savage**; and the fifth is Language.

[1] Lotze, *Microcosmus*, p. 162.

The first source—the Mind of a little child—has just been referred to. Mind, in Man, does not start into being fully ripe. It dawns; it grows; it mellows; it decays. This growing moreover is a gradual growing, an infinitely gentle, never abrupt unfolding—the kind of growing which in every other department of Nature we are taught by Nature to associate with an Evolution. If the Mind of the infant had been evolved, and that not from primeval Man, but from some more ancient animal, it could not to more perfection have simulated the appearance of having so come.

But this is not all. The Mind of a child not only grows, but grows in a certain order. And the astonishing fact about that order is that it is *the probable order of evolution of mental faculty* as a whole. Where Science gets that probable order will be referred to by and by. Meantime, simply note the fact that not only in the manner but in the order of its development, the human Mind simulates a product of Evolution. The Mind of a child, in short, is to be treated as an unfolding embryo; and just as the embryo of the body recapitulates the long life-history of all the bodies that led up to it, so this subtler embryo in running its course through the swift years of early infancy runs up the psychic scale through which, as evidence from another field will show, Mind probably evolved. We have seen also that in the case of the body, each step of progress in the embryo has its equivalent either in the bodies or in the embryos of lower forms of life. Now each phase of mental development in the child is also permanently represented by some species among the lower animals, by idiots, or by the Mind of some existing savage.

9

Let us turn, however, to the second source of information—Mind in the lower Animals.

That animals have " Minds " is a fact which probably no one now disputes. Stories of " Animal Intelligence" and " Animal Sagacity " in dogs and bees and ants and elephants and a hundred other creatures have been told us from childhood with redundant re-iteration. The old protest that animals have no Mind but only instinct has lost its point. In addition to instincts, animals betray intelligence, and often a high degree of intelligence; they share our feelings and emotions; they have memories; they form percepts; they invent new ways of satisfying their desires, they learn by experience. It is true their Minds want much, and all that is highest; but the point is that they actually have Minds, whatever their quantity and whatever their quality.[1] If abstraction, as Locke says, " is an excellency which the faculties of brutes do by no means attain to," we cannot on that account deny them Mind, but only that height of Mind which men have, and which Evolution would never look for in any living thing but Man. An

---

[1] As to the exact point of the difference, Mr. Romanes draws the line at the exclusive possession by Man of the power of introspective reflection in the light of self-consciousness. " Wherein," he asks, " does the distinction truly consist ? It consists in the power which the human being displays of objectifying ideas, or of setting one state of mind before another state, and contemplating the relation between them. The power to think is—or, as I should prefer to state it, the power to think at all—is the power which is given by introspective reflection in the light of self-consciousness. . . . We have no evidence to show that any animal is capable of thus objectifying its own ideas ; and, therefore, we have no evidence that any animal is capable of judg-

Evolutionist would no more expect to find the higher rational characteristics in a wolf or a bear than to unearth the modern turbine from a Roman aqueduct.

Though the possession even of a few rudiments of Mind by animals is a sufficient starting point for Mental Evolution, to say that they have only a few rudiments is to understate the facts. But we know so little what Mind is that speculation in this region can only be done in the rough. On one hand lies the danger of minimizing tremendous distinctions, on the other, of pretending to know all about these distinctions, because we have learned to call them by certain names. Mind, when we come to see what it is, may be one; perhaps must be one. The habit of unconsciously regarding the powers and faculties of Mind as separate entities, like the organs of the body, has its risks as well as its uses; and we cannot too often remind ourselves that this is a mere device to facilitate thought and speech.

It is mainly to Mr. Romanes that we owe the working out of the evidence in this connection; and even though his researches be little more than a preliminary exploration, their general results are striking. Realizing that the most scientific way to discover

ment. Indeed, I will go further and affirm that we have the best evidence which is derivable from what are necessarily ejective sources, to prove that no animal can possibly attain to these excellencies of subjective life." Mr. Romanes proceeds to state the reason why. It is because of " the absence in brutes of the needful conditions to the occurrence of those excellencies as they obtain in themselves . . . the great distinction between the brute and the man really lies behind the faculties both of conception and prediction ; it resides in the conditions to the occurrence of either."—*Mental Evolution in Animals,* p. 175.

whether there are any affinities between Mind in Animals and Mind in Man is to compare the one with the
other, he began a laborious study of the Animal
world. His conclusions are contained in "Animal
Intelligence" and "Mental Evolution in Animals"—
volumes which no one can read without being convinced at least of the thoroughness and fairness of the
investigation. That abundant traces were found of
Mind in the lower animals goes without saying. But
the range of mental phenomena discovered there may
certainly excite surprise. Thus, to consider only one
set of phenomena—that of the emotions—all the following products of emotional development are represented at one stage or another of animal life :

FEAR	EMULATION	BENEVOLENCE
SURPRISE	PRIDE	REVENGE
AFFECTION	RESENTMENT	RAGE
PUGNACITY	EMOTION OF THE	SHAME
CURIOSITY	BEAUTIFUL	REGRET
JEALOUSY	GRIEF	DECEITFULNESS
ANGER	HATE	EMOTION OF THE
PLAY	CRUELTY	LUDICROUS
SYMPATHY		

But this list is something more than a bare catalogue of what human emotions exist in the animal
world. It is an *arranged* catalogue, a more or less
definite psychological scale. These emotions did not
only appear in animals, but they appeared in this
order. Now to find out order in Evolution is of first
importance. For order of events is history, and Evolution is history. In creatures very far down the

scale of life—the Annelids—Mr. Romanes distinguished what appeared to him to be one of the earliest emotions—Fear. Somewhat higher up, among the Insects, he met with the Social Feelings, as well as Industry, Pugnacity, and Curiosity. Jealousy seems to have been born into the world with Fishes; Sympathy with Birds. The Carnivora are responsible for Cruelty, Hate, and Grief; the Anthropoid Apes for Remorse, Shame, the Sense of the Ludicrous, and Deceit.

Now, when we compare this table with a similar table compiled from a careful study of the emotional states in a little child, two striking facts appear. In the first place, there are almost no emotions in the child which are not here—this list, in short, practically exhausts the list of human emotions. With the exception of the religious feelings, the moral sense, and the perception of the sublime, there is nothing found even in adult Man which is not represented with more or less vividness in the Animal Kingdom. But this is not all. These emotions, as already hinted, appear in the Mind of the growing child *in the same order as they appear on the animal scale.* At three weeks, for instance, Fear is perceptibly manifest in a little child. When it is seven weeks old the Social Affections dawn. At twelve weeks emerges Jealousy, with its companion Anger. Sympathy appears after five months; Pride, Resentment, Love of Ornament, after eight; Shame, Remorse, and Sense of the Ludicrous after fifteen. These dates, of course, do not indicate in any mechanical way the birthdays of emotions; they represent rather stages in an infinitely gentle mental ascent, stages nevertheless so

marked that we are able to give them names, and use
them as landmarks in psychogenesis. Yet taken even
as representing a rough order it is a circumstance to
which some significance must be attached that the
tree of Mind as we know it in lower Nature, and the
tree of Mind as we know it in a little child, should be
the same tree, starting its roots at the same place, and
though by no means ending its branches at the same
level, at least growing them so far in a parallel direction.

Do we read these emotions into the lower animals
or are they really there? That they are not there in
the sense in which we think them there is probably
certain. But that they are there in some sense, a
sense sufficient to permit us cautiously to reason
from, seems an admissible hypothesis. No doubt it
takes much for granted,—partly, indeed, the very
thing to be proved. But discounting even the enormous limitations of the inquiry, there is surely a
residuum of general result to make it at least worth
making.

If we turn from emotional to intellectual development, the parallelism though much more faint is at
least shadowed. Again we find a list of intellectual
products common to both Animal and Man, and, again
an approximate order common to both. It is true,
Man's development beyond the highest point attained
by any animal in the region of the intellect, is all but
infinite. Of rational judgment he has the whole monopoly. Wherever the roots of Mind be, there is no
uncertainty as to where, and where exclusively, the
higher branches are. Grant that the mental faculties
of Man and Animal part company at a point, there

remains to consider the vast distance—in the case of the emotions almost the whole distance—where they run parallel with one another. Comparative psychology is not so advanced a science as comparative embryology ; yet no one who has felt the force of the recapitulation argument for the evolution of bodily function, even making all allowances for the differences of the things compared, will deny the weight of the corresponding argument for the evolution of Mind. Why should the Mind thus recapitulate in its development the psychic life of animals unless some vital link connected them ?

A singular complement to this argument has been suggested recently—though as yet only in the form of the vaguest hint—from the side of Mental Pathology. When the Mind is affected by certain diseases, its progress downward can often be followed step by step. It does not tumble down in a moment into chaos like a house of cards, but in a definite order, stone by stone, or story by story. Now the striking thing about that order is, that it is the probable order in which the building has gone up. The order of descent, in short, is the inverse of the order of ascent. The first faculty to go, in many cases of insanity, is the last faculty which arrived ; the next faculty is affected next ; the whole spring uncoiling as it were in the order and direction in which, presumably, it had been wound up. Sometimes even in the phenomenon of old age the cycle may be clearly traced. " Just as consciousness is slowly evolved out of vegetative life, so is it, through the infirmities of old age, the gradual approach of death, and in advanced mental disease, again resolved into it. The highest, most

differentiated phenomena of consciousness are the first
to give way; impulse, instinct, and reflex movements
become again predominant.   The phrase ' to grow
childish' expresses the resemblance between the first
stage and the stage of dissolution." [1]

That the highest part of man should totter first is
what, on the theory of mental evolution, one would
already have expected.   The highest part is the latest
added part, and the latest added part is the least
secured part.   As the last arrival, it is not yet at
home; it has not had time to get lastingly embedded
in the brain; the competition of older faculties is
against it; the hold of the will upon it is slight and
fitful; its tenure as a tenant is precarious and often
threatened.   Among the older and more permanent
residents, therefore, it has little chance.   Hence if
anything goes wrong, as the last added, the most complex, the least automatic of all the functions, it is the
first to suffer.

We are but too familiar with cases where men of
lofty intellect and women of most pure mind, seized
in the awful grasp of madness, are transformed in
a few brief months into beings worse than brutes.
How are we to account, on any other principle than
this, for that most shocking of all catastrophes the
sudden and total break-up, the *devolution*, of a saint?
That the wise man should become a chattering idiot is
inexplicable enough, but that the saintly soul should
riot in blasphemy and immorality so foul that not
among the lowest races is there anything to liken to
it—these are phenomena so staggering that if Evolution hold any key to them at all, its suggestion must

[1] Höffding, *Pyschology*, p. 92.

come as at least a partial relief to the human mind. These are possibly cases of actual reversion, cases where all the beautiful later buildings of humanity had been swept away and only the elemental brute foundations left. Devolution is thus assumed to be a co-relative of Evolution. And as the morbid states of the Mind are more and more studied in this relation, it may yet be possible from the phenomena of insanity to lay bare to some extent the outline of intellectual ascent. In the present state both of psychology, and especially of our knowledge of the brain, nothing probably could be more precarious than this as an argument. The very statement involves modes of expression which exact science would rule out of court. The best that can be said is that it is a suggestion awaiting further light before it can even rank as a theory. Complex as the source of knowledge is, the Mind itself must ever be the final authority on its own biography. Analogy from lower nature may do much to confirm the reading ; the mental history of the human race, from the rudiments of intellect in the savage to its development in civilized life, may contribute some closing chapters ; but unless the Mind tell its own story it will never be fully told. Yet should it ever thus be told, the mystery of Mind itself would remain the same. For the most this could do would be to replace one mystery by a greater. For what greater mystery could there be than that within the mystery of the Mind itself there should lie concealed the very key to unlock its mystery ?

To pass from this fascinating region to the material contributions of Anthropology is a somewhat abrupt

transition.   But this third line of approach to a
knowledge of the earlier phases of Mind need not
detain us long.

So patient has been the search over almost the
whole world for relics of pre-historic Man, that vast
collections are now everywhere available where the
arts, industries, weapons, and, by inference, the mental development, of the earlier inhabitants of this
planet can be practically studied.   On the two main
points at issue in the discussion of mental evolution
these collections are unanimous.   They reveal in the
first instance, traces of Mind of a very low order existing from an unknown antiquity ; and in the second
place, they show a gradual improving of this Mind as
we approach the present day.   It may be that in some
cases the evidence suggests a degenerating rather
than an ascending civilization ; but perturbations of
this sort do not affect the main question, nor neutralize the other facts.   Evolution is constantly confronted with statements as to the former glory of now
decadent nations, as if that were an argument against
the theory.   Granting that nations have degenerated,
it still remains to account for that from which they
degenerated.   That Egypt has fallen from a great
height is certain ; but the real problem is how it got
to that height.   When a boy's kite descends in our
garden, we do not assume that it came from the
clouds.   That it went up before it came down is
obvious from all that we know of kite-making.   And
that nations went up before they came down is obvious from all that we know of nation-making.   The
gravitation, moreover, which brings down nations is
just as real as the gravitation which brings down

kites; and instead of a falling nation being a stumbling block to Evolution, it is a necessity of the theory. The degeneration and extinction of the unfit are as infallibly brought about by natural laws as the survival of the fit. Evolution is by no means synonymous with uninterrupted progress, but at every turn means relapse, extinction, and decay.

It is pretty clear that, applying the old Argument from Design to the case of the most ancient human relics, Man began the Ascent of Civilization at zero. There has been a time in the history of every nation when the only supplements to the organs of the body for the uses of Man were the stones of the field and the sticks of the forest. To use these natural, abundant, and portable objects, was an obvious resource with early tribes. If Mind dawned in the past at all, it is with such objects that we should expect its first associations, and as a matter of fact it seems everywhere to have been so. Relics of a Stick Age would of course be obliterated by time, but traces of a Stone Age have been found, not in connection with the first beginnings of a few tribes only, but with the first beginnings—from the point that any representation is possible—of probably every nation in the world. The wide geographical use of stone implements is one of the most striking facts in Anthropology. Instead of being confined to a few peoples, and to outlying districts, as is sometimes asserted, their distribution is universal. They are found throughout the length and breadth of Europe, and on all its islands; they occur everywhere in Western Asia, and north of the Himalayas. In the Malay Peninsula they strew the ground in endless numbers; and again, in

Australia, New Zealand, New Caledonia, the New Hebrides, and the Coral Islands of the Pacific. Known in China, they are scattered broad-cast throughout Japan, and the same is true of America, Mexico, and Peru. If a child playing with a toy spade is a proof that it is a child, a nation working with stone axes is proved to be a child-nation. Erroneous conclusions may easily be drawn, and indeed have been, from the fact of a nation using stone, but the general law stands. Partly, perhaps, by mutual intercourse, this use of stone became universal; but it arose, more likely, from the similarity in primitive needs, and the available means of gratifying them. Living under widely different conditions, and in every variety of climate, all early peoples shared the instincts of humanity which first called in the use of tools and weapons. All felt the same hunger; all had the instinct of self-preservation; and the universality of these instincts and the commonness of stone led the groping Mind to fasten upon it, and make it one of the first steps to the Arts. A Stone Age, thus, was the natural beginning. In the nature of things there could have been no earlier. If Mind really grew by infinitely gradual ascents, the exact situation the theory requires is here provided in actual fact.

The next step from the Stone Age, so far as further appeal to ancient implements can guide us, is also exactly what one would expect. It is to a *better* Stone Age. Two distinct grades of stone implements are found, the rough and the smooth, or the unground and the ground. For a long period the idea never seems to have dawned that a smooth stone made a better axe than a rough one. Mind was as yet unequal to this small discovery, and there are vast remains representing long intervals of time where all the stone implements and tools are of the unground type. Even when the hour did come, when savage vied with savage in putting the finest polish on his flints, his inspiration probably came from Nature. The first lapidary was the sea; the smoothed pebble on the beach, or the rounded stone of the mountain stream, supplied the pattern. There is no question that the rough stone came earlier than the ground stone. Thus the implements of the Drift Period, those of the Danish Mounds, the Bone Caves, and the gravels of St. Acheul are mostly unground, while those of the later Lake-Dwellers are almost wholly of the smooth type.

To follow the Stone Age upward into the Bronze Period, and from that to the Age of Iron is not necessary for the present purpose. For at this point the order of succession passes from shell-mound and crannog, into living hands. There are nations with us still who have climbed so short a distance up the psychic scale as to be still in the Age of Stone— peoples whose mental culture and habits are often actual witnesses to the mental states of early Man. These children of Nature take up the thread of mental progress where the Troglodyte and Drift-Man left it; and the modern traveller, starting from the civilization of Europe can follow Mind downwards step by step, in ever descending order, tracing its shadings backwards to a first simplicity till he finds himself with the still living Lake-dweller of Nyasaland or the Bushman of the African forest. Time was when these humble tribes, with their strange and artless ways, were mere food for the curious. Now the study of the lower native races has risen to the first rank in comparative psychology; and the student of beginnings, whether they be the beginnings of Art or of Ethics, of Language or of Letters, of Law or of Religion, goes to seek the roots of his science in the ways, traditions, faiths, and institutions of savage life.

This leads us, however, to the fourth of the sources from which we were to gather a hint or two with regard to the past of Mind—the savage. No one should pronounce upon the Evolution of Mind till he has seen a savage. By this is not meant the show savage of an Australian town, or the quay Kaffir of a South African port, or the Reservation Indian of a Western State; but the savage as he is in reality, and as he may be seen to-day by any who care to look upon so weird a spectacle. No study from the life can compare with this in interest or in pathos, nor stir so many strange emotions in the mind of a thoughtful man. To sit with this incalculable creature in the heart of the great forest; to live with him in his natural home as the guest of Nature, to watch his ways and moods and try to resolve the ceaseless mystery of his thoughts—this, whether the existing savage represents the primitive savage or not, is to open one of the workshops of Creation and behold the half-finished product from which humanity has been evolved.

The world is getting old, but the traveller who cares to follow the daybreak of Mind for himself can almost do so still. Selecting a region where the wand of western civilization has scarcely reached, let him begin with a cruise in the Malay Archipelago or in the Coral Seas of the Southern Pacific. He may find himself there even yet on spots on which no white foot has ever trod, on islands where unknown races have worked out their destiny for untold centuries, whose teeming peoples have no name, and whose habits and mode of life are only known to the outer world through a ship's telescope. As he coasts along, he will see the dusky figures steal like shades among the trees, or hurry past in their bark canoes, or crouch in fear upon the coral sand. He can watch them gather the bread-fruit from the tree and pull the cocoa-nut from the palm and root out the taro for a meal which, all the year round and all the centuries through, has never changed. In an hour or two he can compass almost the whole round of their simple life, and realize the gulf between himself and them in at least one way—in the utter impossibility of framing to himself an image of the mental world of men and women whose only world is this.

Let him pass on to the coast of Northern Queensland, and, landing where fear of the white man makes landing possible, penetrate the Australian bush. Though the settlements of the European have been there for a generation, he will find the child of Nature still untouched, and neither by intercourse nor imitation removed by one degree from the lowest savage state. These aboriginal peoples know neither house nor home. They neither sow nor reap. Their weapons are those of Nature, a pointed stick and a knotted club. They live like wild things on roots and berries and birds and wallabies, and

in the monotony of their life and the uncouthness of
their Mind represent almost the lowest level of humanity.[1]

From these rudiments of mankind let him make his
way to the New Hebrides, to Tana, and Santo, and
Ambrym, and Aurora. These islands, besides Man,
contain only three things, coral, lava, and trees. Until but yesterday their peoples had never seen anything but coral, lava, and trees. They did not know
that there was anything else in the world. One hundred years ago Captain Cook discovered these islanders and gave them a few nails. They planted them in
the ground that they might grow into bigger nails.
It is true that in other lands a very rich life and a
very wide world could be made out of no more varied
materials than coral, lava and trees; but on these
Tropical Islands Nature is disastrously kind. All
that her children need is provided for them ready-
made. Her sun shines on them so that they are never
either cold or hot; she provides crops for them in un-
exampled luxuriance, and arranges the year to be one
long harvest; she allows no wild animals to prowl
among the forest; and surrounding them with the
alienating sea she preserves them from the attacks of
human enemies. Outside the struggle for life, they
are out of life itself. Treated as children, they remain children. To look at them now is to recall the

---

[1] The situation is dramatic, that from end to end of the region
occupied by these tribes, there stretches the Telegraph connecting Australia with Europe. But what is at once dramatic and
pathetic is that the natives know it only in its material relations
—as so much wire, the first metal they have ever seen, to cut
into lengths for spear-heads.

long holiday of the childhood of the world. It is to behold one's natural face in a glass.

Pass on through the other Cannibal Islands and, apart from the improvement of weapons and the construction of a hut, throughout vast regions there is still no sign of mental progress. But before one has completed the circuit of the Pacific the change begins to come. Gradually there appear the beginnings of industry and even of art. In the Solomon Group and in New Guinea, carving and painting may be seen in an early infancy. The canoes are large and good, fish-hooks are manufactured and weaving of a rude kind has been established. There can be no question at this stage that the Mind of Man has begun its upward path. And what now begins to impress one is not the poverty of the early Mind, but the enormous potentialities that lie within it, and the exceeding swiftness of its Ascent towards higher things. When the Sandwich Islands are reached, the contrast appears in its full significance. Here, a century ago, Captain Cook, through whom the first knowledge of their existence reached the outer world, was killed and eaten. To-day the children of his murderers have taken their place among the civilized nations of the world, and their Kings and Queens demand acknowledgment at modern Courts.

Books have been given to the world on the Mind of animals. It is strange that so little should have been written specifically on the Mind of the savage. But though this living mine has not yet been drawn upon for its last contribution to science, facts to suggest and sustain a theory of mental evolution are everywhere abundant. Waiving individual cases where

10

nations have fallen from a higher intellectual level the. proof indicates a rising potentiality and widening of range as we pass from primitive to civilized states. It is open to debate whether during the historic period mere intellectual advance has been considerable, whether more penetrating or commanding intellects have ever appeared than those of Job, Isaiah, Plato, Shakspeare. But that is matter of yesterday. What concerns us now to note is that the Mind of Man as a whole has had a slow and gradual dawn; that it has existed, and exists to-day, among certain tribes at almost the lowest point of development with which the word human can be associated; and that from that point an Ascent of Mind can be traced from tribe to nation in an ever increasing complexity and through infinitely delicate shades of improvement, till the highest civilized states are reached. In the very nature of things we should have expected such a result. For this is not only a question of faculty. In a far more intimate sense than we are apt to imagine, it is a question of a gradually evolving environment. Every infinitesimal enrichment of the soil for Mind to grow in meant an infinitesimal enrichment of the Mind itself. " It needs but to ask what would happen to ourselves were the whole mass of existing knowledge obliterated, and were children with nothing beyond their nursery-language left to grow up without guidance or instruction from adults, to perceive that even now the higher intellectual faculties would be almost inoperative, from lack of the materials and aids accumulated by past civilization. And seeing this, we cannot fail to see that development of the higher intellectual faculties has gone on *pari passu* with social advance alike as cause and consequence; that the primitive man could not evolve these higher intellectual faculties in the absence of a fit environment; and that in this, as in other respects, his progress was retarded by the absence of capacities which only progress could bring." [1]

The last testimony is that of Language. It has already been pleaded in excuse for the absence of actual proof for mental evolution that Mind leaves no material footprints by which the palæontologist can trace its upward path. Yet this is not wholly true. The flints and arrow-heads, the celts and hammers, of early Man are fossil intelligence; the remains of primitive arts and industries are petrified Mind. But there is one mould into which Mind has run more large and beautiful than any of these. When its contents are examined they carry us back not only to what men worked at with their hands, but to what they said to one another as they worked and what they thought as they spoke. That mould is Language. Language, says Jean Paul, is " ein Wörter-buch erblasster Metaphern "—a dictionary of faded metaphors. But it is much more. A word is a counter of the brain, a tangible expression of a mental state, an heirloom of the wealth of culture of a race. And an old word, like an ancient coin, speaks to us of a former currency of thought, and by its image and superscription reveals the mental life and aspiration of those who minted it. " Language is the amber in which a thousand precious and subtle thoughts have been safely embalmed and preserved. It is the embodiment, the incarnation, of the feelings and thoughts

[1] Herbert Spencer, *Principles of Sociology*, Vol. I., p. 90, 1.

and experiences of a nation, yea often of many nations, and of all which through long centuries they have attained to and won. It stands like the Pillars of Hercules, to mark how far the moral and intellectual conquests of mankind have advanced, only not like those pillars, fixed and immovable, but even itself advancing with the progress of these. The mighty moral instincts which have been working in the popular mind have found therein their unconscious voice ; and the single kinglier spirits that have looked deeper into the heart of things have oftentimes gathered up all they have seen into some one word, which they have launched upon the world, and with which they have enriched it forever—making in the new word a new region of thought to be henceforward in some sort the common heritage of all." [1]

What then, when we open this marvellous structure, is the revelation yielded us of the mental states of those who lived at the dawn of speech? An impression of poverty, great and pathetic. All fossils teach the same lesson—the lesson of life, beauty, structure, waning into a poverty-stricken past. Whether they be the shells which living creatures once inhabited, or the bones of departed vertebrate types, or the forms of words where wisdom lay entombed, the structures became simpler and simpler cruder and cruder, less full of the richness and abundance of life as we near the birth of time. They tell of days when the world was very young, when plants were flowerless and animals back-boneless, of later years when primeval Man prowled the forest and chipped his flints and chattered in uncouth syllables

[1] Trench, *The Study of Words,* p. 28.

of battle and the chase. No words entered at that time into human speech except those relating to the activities, few and monotonous, of an almost animal lot. These were the days of the protoplasm of speech. There was no differentiation between verbs or adverbs, nouns or adjectives. The sentence as yet was not; each word was a sentence. There was no grammatical inflection but the inflection of the voice; the moods of the verb were uttered by intonation or grimace. The pronouns " him " and " you " were made by pointing at him and you. Man had even no word for himself, for he had not yet discovered himself. This fact, when duly considered, raises the witness of Language to the Ascent of Mind to an almost unique importance. Nothing more significant could be said as to Man's mental past than that there was a time when he was scarcely conscious of himself, as a self. He knew himself, not as subject, but like a little child, as one of the objects of the external world. The words might have been written historically of mankind, " When I was a child, I spake as a child."

This evidence will meet us again in other forms when we pass to consider the Evolution of Language itself. Meantime let us close this chapter by pointing out a relation of a much more significant order between Language and the whole subject of Mental Evolution. For the point is not only of special interest but it touches upon, and helps to solve, one of the vital problems of the Ascent of Man.

The enormous distance travelled by the Mind of Man beyond the utmost limit of intelligence reached by any animal is a puzzling circumstance, a circum-

stance only equalled in strangeness by another—
the suddenness with which that rise took place. Both
facts are without a parallel in nature. Why, of the
countless thousands of species of animals, each with
some shadowy rudiment of a Mind, all should have
remained comparatively at the same dead level,
while Man alone shot past and developed powers of
a quality and with a speed unknown in the world's
history, is a question which it is impossible not to
raise. That by far the greatest step in the world's
history should not only have been taken at the
eleventh hour, but that it took only an hour to do
it—for compared with the time when animals began
their first activities, the birth of Man is a thing of
yesterday—seems almost the denial of Evolution.
What was it in Man's case that gave his mental
powers their unprecedented start or facilitated a
growth so rapid and so vast?

The factors in all Evolution, and above all in this,
are too subtle to encourage one to speculate with
final assurance on so fine a problem. Nevertheless,
when it is asked, What brought about this sudden
rise of intelligence in the case of Man, there is a
wonderful unanimity among men of science as to the
answer. It came about, it is supposed, in connection
with the acquisition by Man of the power to express
his mind, that is to *speak*. Evolution, up to this time,
had only one way of banking the gains it won—heredity. To hand on any improvement physically was
a slow and precarious work. But with the discovery
of language there arose a new method of passing on a
step in progress. Instead of sowing the gain on the
wind of heredity, it was fastened on the wings of

words. The way to make money is not only to accumulate small gains steadily, but to put them out at
a good rate of interest. Animals did the first with
their mental acquisitions : Man did the second. At a
comparatively early date, he found out a first-rate
and permanent investment for his money, so that he
could not only keep his savings and put them out at
the highest rate of interest, but have a share in all
the gain that was made by other men. That discovery was Language. Many animals had hit upon
an imperfect form of this discovery ; but Man alone
succeeded in improving it up to a really paying point.
The condition of all growth is exercise, and till he
could find a further field and a larger opportunity to
work what little brains he had, he had little chance
of getting more. Speech gave him this opportunity.
He rapidly ran up a fortune in brain-matter, because
he had found out new uses for it, new exercises of it,
and especially a permanent investment for husbanding
in the race each gain as it was made in the individual.
When he did anything he could now *say* it; when he
learned anything he could pass it on ; when he became
wise wisdom did not die with him, it was banked in
the Mind of humanity. So one man lent his mind to
another. The loans became larger and larger, the
interest greater and greater ; Man's fortune was
secured. In the mere Struggle for Life, his wits were
sharpened up to a point; but unless he had learned to
talk, he could never have passed very far beyond the
animal.

Apart from the saving of time and the facility for
increased knowledge, the acquisition of speech meant
a saving of brain. A word is a counter for a thought.

To use language is to make thinking easy. Hence the release of brain energy for further developments in new directions. In these and other ways speech became the main factor in the intellectual development of mankind. Language formed the trellis on which Mind climbed upward, which continuously sustained the ripening fruits of knowledge for later minds to pluck. Before the savage's son was ten years old he knew all that his father knew. The ways of the game, the habits of birds and fish, the construction of traps and snares—all these would be taught him. The physical world, the changes of season, the location of hostile tribes, the strategies of war, all the details and interests of savage life would be explained. And before the boy was in his teens he was equipped for the Struggle for Life as his forefathers had never been even in old age. The son, in short, started to evolve where his father left off. Try to realize what it would be for each of us to begin life afresh, to be able to learn nothing by the experiences of others, to live in a dumb and illiterate world, and see what chance the animal had of making pronounced progress until the acquisition of speech. It is not too much to say that speech, if mental evolution is to come to anything or is to be worth anything, is a necessary condition. By it alone, in any degree worth naming, can the fruits of observation and experience of one generation be husbanded to form a new starting-point for a second, nor without it could there be any concerted action or social life. The greatness of the human Mind, after all, is due to the tongue, the material instrument of reason, and to Language the outward expression of the inner life.

CHAPTER V.

THE EVOLUTION OF LANGUAGE.

If Evolution is the method of Creation, the faculty of Speech was no sudden gift. Man's mind is not to be thought of as the cylinder of a phonograph to which ready-made words were spoken and stored up for future use. Before *Homo sapiens* was evolved he must necessarily have been preceded for a longer or shorter period by *Homo alalus*, the not-speaking man; and this man had to make his words, and beginning with dumb signs and inarticulate cries to build up a body of Language word by word as the body was built up cell by cell.

The alternative theory of the origin of Language universally held until lately, and expressed in so many words even by the eighth edition of the *Encyclopædia Britannica*, that "our first parents received it by immediate inspiration," has the same relation to exact science as the view that the world was made in six days by direct creative fiat. Both are poetically true. But to science, seeking for precise methods of operation, neither is an adequate statement of now ascertained facts. The same processes of research that made the poetic view of creation untenable in the physical realm are now slowly beginning

153

to displace the older view of the origin of speech.
That Language should be outside a law whose universality is being established with every step of progress,
is itself improbable; and now that the field is being
exhaustively explored the proofs that it is no exception multiply on every side. The living interest the
mere suggestion gives to the study of Language is
obvious. Evolution enters no region—dull, neglected,
or remote—of the temple of knowledge without transforming it. Philology, since this wizard touched it,
has become one of the most entrancing of the sciences.
And Language, from a study which interested only a
few specialists, is disclosed as one vast palimpsest,
every word and phrase luminous with the inner mind
and soul of the past. To penetrate far into this
tempting region is beyond our province now. The
immediate object is to give a simple sketch of the
possible conditions which first led Man to speak; of
the principles which apparently guided the formation
of his early vocabulary ; and of the gradual refining of
the means of intercommunication between him and his
fellow-men as time passed on. Instead of beginning
with words, therefore, we shall begin with Man. For
the first condition for understanding the Evolution of
Speech is that we take it up as a study from the life,
that we place ourselves in the primeval forest with
early Man, in touch with the actual scenes in which
he lived, and note the real experiences and necessities
of such a lot. We may indeed discover in this research small trace of a miraculous inbreathing of
formal words. But to make Speech and fit it into
a man, after all is said, is less miraculous than to fit a
man to make Speech.

One of the earliest devices hit upon in the course of Evolution was the principle of co-operation. Long before men had learned to form themselves into tribes and clans for mutual strength and service, gregariousness was an established institution. The deer had formed themselves into herds, and the monkeys into troops; the birds were in flocks, and the wolves in packs; the bees in hives, and the ants in colonies. And so abundant and dominant in every part of the world are these social types to-day that we may be sure the gregarious state has exceptional advantages in the upward struggle.

One of these advantages, obviously, is the mere physical strength of numbers. But there is another and a much more important one—the mental strength of a combination. Here is a herd of deer, scattered, as they love to be, in a string, quarter of a mile long. Every animal in the herd not only shares the physical strength of all the rest, but their powers of observation. Its foresight in presence of possible danger is the foresight of the herd. It has as many eyes as the herd, as many ears, as many organs of smell, its nervous system extends throughout the whole space covered by the line; its environment, in short, is not only what it hears, sees, smells, touches, tastes, but what every single member hears, sees, smells, touches, tastes. This means an enormous advantage in the Struggle for Life. What deer have to arm themselves most against is surprise. When it comes to an actual fight, comrades are of little use. At that crisis the others run away and leave the victims to their fate. But in helping one another to avert that crisis, the **value of this mutual aid is so great that gregarious**

animals, for the most part timid and defenceless as
individuals, have survived to occupy in untold multitudes the highest places in Nature.

The success of the co-operative principle, however,
depends upon one condition : the members of the herd
must be able to communicate with one another. It
matters not how acute the senses of each animal may
be, the strength of the column depends on the power
to transmit from one to another what impressions
each may receive at any moment from without.
Without this power the sociality of the herd is stultified ; the army, having no signalling department, is
powerless as an army. But if any member of the
herd is able by motion of head or foot or neck or ear,
by any sign or by any sound, to pass on the news that
there is danger near, each instantly enters into possession of the faculties of the whole. Each has a hundred eyes, noses, ears. Each has quarter of a mile
of nerves. Thus numbers are strength only when
strength is coupled with some power of intercommunication by signs. If one herd develops this signalling system and another does not, its chances of
survival will be greater. The less equipped herds will
be slowly decimated and driven to the wall; and
those which survive to propagate their kind will be
those whose signal-service is most efficient and complete. Hence the Evolution of the signal-system.
Under the influence of Natural Selection its progress
was inevitable. New circumstances and relations
would in time arise, calling for additions, vocal, visible, audible, to the sign-vocabulary. And as time
went on each set of animals would acquire a definite
signal-service of its own, elementary to the last

degree, yet covering the range of its ordinary experiences and adequate to the expression of its limited mental states.

Now what interests us with regard to these signs is that they are *Language.* The evolution we have been tracing is nothing less than the first stage in the evolution of Speech. Any means by which information is conveyed from one mind to another is Language. And Language existed on the earth from the day that animals began to live together. The mere fact that animals cling to one another, live together, move about together, proves that they communicate. Among the ants, perhaps the most social of the lower animals, this power is so perfect that they are not merely endowed with a few general signs but seem able to convey information upon matters of detail. Sweeping across country in great armies they keep up communication throughout the whole line, and succeed in conveying to one another information as to the easiest routes, the presence of enemies or obstacles, the proximity of food supplies, and even of the numbers required on emergencies to leave the main band for any special service. Every one has observed ants stop when they meet one another and exchange a rapid greeting by means of their waving antennæ, and it is possibly through these perplexing organs that definite intercourse between one creature and another first entered the world. The exact nature of the antenna-language is not yet fathomed, but the perfection to which it is carried proves that the idea of language generally has existed in nature from the earliest time. Among higher animals various outward expressions of emotions are made, and these become of service in time for the conveyance of information to others. The howl of the dog, the neigh of the horse, the bleat of the lamb, the stamp of the goat, and other signs are all readily understood by other animals. One monkey utters at least six different sounds to express its feelings; and Mr. Darwin has detected four or five modulations in the bark of the dog: "the bark of eagerness, as in the chase; that of anger as well as growling; the yelp or howl of despair when shut up; the baying at night; the bark of joy when starting on a walk with his master; and the very distinct one of demand or supplication, as when wishing for a door or window to be opened." [1]

Now these signs are as much language as spoken words. You have only to evolve this to get all the language the dictionary-maker requires. Any method of communication, as already said, is Language, and to understand Language we must fix in our minds the idea that it has no necessary connection with actual words. In the simple instances just given there are illustrations of at least three kinds of Language. When a deer throws up its head suddenly, all the other deer throw up their heads. That is a sign. It means "listen." If the first deer sees the object, which has called its attention, to be suspicious, it utters a low note. That is a word. It means "caution." If next it sees the object to be not only suspicious but dangerous, it makes a further use of Language—intonation. Instead of the low note "listen," it utters a sharp loud cry that means "Run for your life." Hence these three kinds of

[1] Darwin, *Descent of Man*, p. 84.

Language—a sign or gesture, a note or word, an intonaticn.

Down to this present hour these are still the three great kinds of Language. The movement of foot or ear has been evolved into the modern gesture or grimace; the note or cry into a word, and the intonation into an emphasis or inflection of the voice. These are still, indeed, not only the main elements in Language but the only elements. The eloquence which enthralls the legislators of St. Stephen's, or the appeal which melts the worshippers at St. Paul's, originated in the voices of the forest and the activities of the ant-hill. To those who have not realized the exceeding smallness of the beginnings of all new developments, the suggestion of science as to the origin of Language, like many of its other suggestions about early stages, will seem almost ludicrous. But a knowledge of two things warns one not to look for surprises at the beginning of Evolution but at the end. In the first place, it is all but a cardinal principle that developments are brought about by minute, slow and insensible degrees. The second fact is even more important. The theatre of change is the actual world, and the exciting cause something really happening in every-day life. New departures are not made in the air. They arise in connection with some commonplace event; and usually take the shape of some slightly new response. In other connections, of course, the converse is also true, but when a change occurs for the first time in the life of an organism the exciting cause, whatever the internal adaptation, or want of it, is some change in the environment. Among the events then, actually happening in the day's round, we are to seek for the exciting cause of the earliest forms of speech.

The simplest Language open to Man was that which we have already seen to mark the beginning of all Language, the Language of gesture or sign. To the word gesture, however, it is necessary to attach a larger meaning than the term ordinarily expresses to us. It is not to be limited, for example, to visible movements of the limbs or facial muscles. The ejaculations of the savage, the drumming of the gorilla, the screech of the parrot, the crying, growling, purring, hissing, and spitting of other animals are all forms of gesture. Nor is it possible to separate the Language of gesture from the Language of intonation. These have grown up side by side and can neither be distinguished psychologically nor as to priority in the order of Evolution. Intonation, though it has grown to be infinitely the more delicate instrument of the two and is still so important a part of some Languages— the Chinese, for example—as to be an integral part of them, has its roots in the same soil and must be looked upon as, along with it, the earliest form of Language.

That this Gesture-Language marked, if not the dawn, at least a very early stage of Language in the case of Man, there is abundant evidence. Apart from analogy, there are at least three witnesses who may be cited in proof not only of the fact, but of the high perfection to which a Gesture-Language may be carried. The first of these witnesses is the *homo alalus*, the not-speaking man, of to-day, the deaf mute. As an actual case of a human being reduced as regards the power of speech to the level of early Man his evidence, even with all allowances for the high development of his mental faculties,

is of scientific value. The mere fact that a deaf man is also a dumb man is almost a final answer to the affirmation that the power of speech is an original and intuitive faculty of Man. If it were so, there is no reason why a deaf man should not speak. The vocal apparatus in his case is complete; all that is required to make him utter a definite sound is to hear one. When he hears one, but not till then, he can imitate it. Language, so far as the testimony of the deaf-mute goes, is clearly a matter of imitation. Unable to attain the second stage of Language—words—he has to content himself with the first—signs. And this Language he has evolved to its last perfection. It shows how little the mere utterance of words has to do with Language, that the deaf-mute is able to converse on every-day subjects almost as perfectly as those who can speak. The permutations and combinations that can be produced with ten pliable fingers, or with the varying expressions of the muscles of the face, are endless, and everything that he cares to know can be uttered or translated to him by motion, gesture, and grimace. To give an idea how far gestures can be made to do the work of spoken words, the signs may be described in which a deaf-and-dumb man once told a child's story in presence of Mr. Tylor. " He began by moving his hand, palm down, about a yard from the ground, as we do to show the height of a child—this meant that it was a child he was thinking of. Then he tied an imaginary pair of bonnet-strings under his chin (his usual sign for female), to make it understood that the child was a little girl. The child's mother was then brought on the scene in a similar way. She beckons to the child and gives her twopence, these being indicated
11

by pretending to drop two coins from one hand into the other; if there had been any doubt as to whether they were copper or silver coins, this would have been settled by pointing to something brown or even by one's contemptuous way of handling coppers which at once distinguishes them from silver. The mother also gives the child a jar, shown by sketching its shape with the forefingers in the air, and going through the act of handing it over. Then by imitating the unmistakable kind of twist with which one turns a treacle-spoon, it is made known that it is treacle the child has to buy. Next, a wave of the hand shows the child being sent off on her errand, the usual sign of walking being added, which is made by two fingers walking on the table. The turning of an imaginary door-handle now takes us into the shop, when the counter is shown by passing the flat hands as it were over it. Behind this counter a figure is pointed out; he is shown to be a man by the usual sign of putting one's hand to one's chin and drawing it down where the beard is or would be; then the sign of tying an apron around one's waist adds the information that the man is the shopman. To him the child gives her jar, dropping the money into his hand, and moving her forefinger as if taking up treacle to show what she wants. Then we see the jar put into an imaginary pair of scales which go up and down; the great treacle-jar is brought from the shelf and the little one filled, with the proper twist to take up the last trickling thread; the grocer puts the two coins in the till, and the little girl sets off with the jar. The deaf-and-dumb story-teller went on to show in pantomime how the child, looking down at the jar, saw a drop of treacle on the rim, wiped it off with her finger, and put the finger in her mouth, how she was tempted to take more, how her mother found her out by the spot of treacle on her pinafore, and so forth." [1]

A second witness is savage Man. Some of the more primitive races, far as they have evolved past the *alalus* stage, still cling to the gesture-language which bulked so largely in the intercourse of their ancestors. No one who has witnessed a conversation—one says " witnessed," for it is more seeing than hearing—between two different tribes of Indians can have any doubt of the working efficiency of this method of speech. After ten minutes of almost pure pantomime each will have told the other everything that it is needful to say. Indians of different tribes, indeed, are able to communicate most perfectly on all ordinary subjects with no more use of the voice than that required for the emission of a few different kinds of grunts. The fact that stranger tribes make so large a use of gesture in expressing themselves to one another does not, of course, imply that each has not a word-language of its own. But few of the Languages of primitive peoples are complete without the additions which gesture offers. There are gaps in the vocabulary of almost all savage tribes due to the fact that in actual speech the *lacunæ* are bridged by signs, and many of their words belong more to the category of signs than to that of words.

The final witness is the first attempt at Language of a little child. Universally an infant opens communication with the mental world around it in the primitive language of gesture and tone. Long before it has learned to speak, without the use of a single word it

[1] Tylor, *Anthropology.*

conveys information as to fundamental wants, and expresses all its varying moods and wishes with a vehemence and point which are almost the envy of riper years. The interesting thing about this is that it is spontaneous. In later childhood it has to be *taught* to speak—because speech is a fine art—but to utter the hereditary and primitive Language of mankind requires no prompting. Words are conventional, movements and sounds are natural. The Language of the nursery is the native Language of the forest, the inarticulate cry of the animal, the intonation of the savage. To quote from Mallery :—" The wishes and emotions of very young children are conveyed in a small number of sounds, but in a great variety of gestures and facial expressions. A child's gestures are intelligent long in advance of speech ; although very early and persistent attempts are made to give it instruction in the latter but none in the former, from the time when it begins *risus cognoscere matrem*. It learns words only as they are taught, and learns them through the medium of signs which are not expressly taught. Long after familiarity with speech it consults the gestures and facial expressions of its parents and nurses, as if seeking them to translate or explain their words. These facts are important in reference to the biologic law that the order of development of the individual is the same as that of the species. . . . The insane understand and obey gestures when they have no knowledge whatever of words. It is also found that semi-idiotic children who cannot be taught more than the merest rudiment of speech can receive a considerable amount of information through signs, and can express themselves by them. Sufferers from aphasia continue to use appropriate gestures. A stammerer, too, works his arms and features as if determined to get his thoughts out, in a manner not only suggestive of the physical struggle, but of the use of gesture as a hereditary expedient." [1]

The survival both of gesture and intonation in modern adult speech, and especially the unconsciousness of their use, illustrate how indelibly these primitive forms of Language are embedded in the human race. There are doubtless exceptions, but it is probably the rule that gestures are mainly called in to supplement expression when the subject-matter of discourse does not belong to the highest ranges of thought, or the speaker to the loftiest type of oratory. The higher levels of thought were reached when the purer forms of spoken Language had become the vehicle of expression; and, as has often been noticed, when a speaker soars into a very lofty region, or allows his mind to grapple intensely and absorbingly with an exalted theme, he becomes more and more motionless, and only resumes the gesture-language when he descends to commoner levels. It is not only that a fine speaker has a greater command of words and is able to dispense with auxiliaries—as a master of style can dispense with the use of italics—but that, at all events, in the case of abstract thought, it is untranslatable into gesture-speech. Gestures are suggestions and reminders of things seen and heard. They are nearly all attached to objects or to moods, and rival words only when used of every-day things.

---

[1] *First Annual Report of the Bureau of Ethnology*, Washington, 1881.

"No sign talker," Mr. Romanes reminds us, "with any amount of time at his disposal, could translate into the language of gesture a page of Kant." [1]

The next stage in the Evolution of Language must have been reached as naturally as the Language of gesture and tone. From the gesture-language to mixtures of signs and sounds, and finally to the specialization of sound into words, is a necessary transition. Apart from the fact that gestures and tones have limits, circumstances must often have arisen in the life of early Man when gesture was impossible. A sign Language is of no use when one savage is at one end of a wood and his wife at the other. He must now roar; and to make his roar explicit, he must have a vocabulary of roars, and of all shades of roars. In the darkness of night also, his signs are useless, and he must now whisper and have a vocabulary of whispers. Nor is it difficult to conceive where he got his first brief list of words. Instead of drawing things in the air with his finger, he would now try to imitate their sounds. Everything around him that conveyed any impression of sound would have associated with it some self-expressive word, which all familiar with the original sound could instantly recognize. Imagine, for instance, a herd of buffalo browsing in a glade of the African forest. The vanguard, some little distance from its neighbors, hears the low growl of a lion. That growl, of course, is Language, and the buffalo understands it as well as we do when the word "lion" is pronounced. Between the word "lion" spoken, and the object lion growled, there is no difference in

[1] *Mental Evolution,* p. 147.

the effect. Suppose, next, the buffalo wished to convey to its comrades the knowledge that a lion was
near, a lion and not some other animal, it might
imitate this growl. It is not likely that it would do
so; some other sign expressing alarm in general
would probably be used, for the discrimination of the
different sources of danger is probably an achievement beyond this animal's power. But if Primitive
Man was placed under the same circumstances, granting that he had begun in a feeble way to exercise
mind, he would almost certainly come in time to
denote a lion by an imitated growl, a wolf by an
imitated whine, and so on. The sighing of the wind,
the flowing of the stream, the beat of the surf, the
note of the bird, the chirp of the grasshopper, the hiss
of the snake, would each be used to express these
things. And gradually a Language would be built up
which included all the things in the environment with
which sound was either directly, indirectly, or accidentally associated.

That this method of word-making is natural is seen
in the facility with which it is still used by children;
and from the early age at which they begin to employ
it, the sound Language is clearly one of the very first
forms of speech. All a child's words are of course
gathered through the sense of hearing, but if it can
itself pick up a word direct from the object, it will use
it long before it elects to repeat the conventional
name taught it by its nurse. The child who says *moo*
for cow, or *bow-wow* for dog, or *tick-tick* for watch, or
puff-puff for train, is an authority on the origin of
human speech. Its father, when he talks of the *hum*
of machinery or the *boom* of the cannon, when he calls

champagne *fizz* or a less aristocratic beverage *pop*, is following in the wake of the inventors of Language. Among savage peoples, and especially those encountering the first rush of new things and thoughts brought them by the advancing wave of civilization, word-making is still going on ; and wherever possible the favorite principle seems to be that of sound.[1]

How full all Languages are of these sound-words is known to the philologist, though multitudes of words in every Language have had their pedigree effaced or obscured by time. " An Englishman would hardly guess from the present pronunciation and meaning of the word pipe what its origin was ; yet when he compares it with the Low Latin pipa, French pipe, pronounced more like our word peep, to chirp, and meaning such a reed-pipe as shepherds played on, he then sees how cleverly the very sound of the musical pipe has been made into a word for all kinds of tubes, such as tobacco-pipes and water-pipes. Words like this travel like Indians on the war-path, wiping out their footmarks as they go. For all we know multi-

---

[1] Among the Coral Islands of the Pacific the savages everywhere speak of the white residents in New Caledonia as the *Wee-wee men,* or *Wee-wees.* Cannibals on a dozen different islands, speaking as many languages, have all this name in common. New Caledonia is a French Penal Settlement, containing thousands of French convicts, and one's first crude thought is that the Wee-wees are so named from their size. A moment's reflection, however, shows that it is taken from their *sounds*—that in fact we have here a very pretty example of modern onomatopœia. These convicts, freed or escaped, find their way over the Pacific group ; and the natives, seizing at once upon their characteristic sound, know them as *Oui-oui's*—a name which has now become general for all Frenchmen in the Southern Pacific.

tudes of our ordinary words may have thus been made from real sounds, but have now lost beyond recovery the traces of their first expressiveness." [1] In the Chinuk language of the West Coast of America, to cite a few more of Tylor's instances, a tavern is called a "*heehee-house*," that is a laughter house, or an amusement house, the word for amusement being taken by an obvious association from the laughter which it excites. How indirect a derivation may be is illustrated by the word which the Basutos of South Africa use for *courtier*. The buzz of a certain fly resembles the sound *ntsi-ntsi*, and they apply this word to those who buzz round the chief as a fly buzzes round a piece of meat. As every one knows " papa " for father, is evolved into papa the pope, and " abba " the Hebrew for father into abbot. For plurals, a doubling of the word is often used, but no doubt at first quantity was expressed by gestures or by numbering on the fingers. " Orang " is the Malay for Man, " Orang-orang " for men while " Orang-utan " is wild man. Verbs are formed on the same principle as nouns. In the Tecuna language of Brazil the verb to sneeze is *haitschu*, while the Welsh for a sneeze is *tis*. Other verbs which came to have large and comprehensive meanings arose out of the simple activities and occupations of primitive life. Thus the first verb in the Bible, the Hebrew " bara " now meaning create, was originally used for cutting or hewing, the first step in making things. In the Borneo language of Africa, the verb " to make " comes from the word *tando*, to weave. In English, " to suffer " meant to bear as a burden, and to " apprehend an idea " was originally to " catch

[1] Tylor, *Anthropology*, p. 127.

hold " of some " sight." Even Max Müller who opposes the onomatopoetic theory with regard to the
origin of most words, agrees that the sounds of the
occupation of men, and especially of men working
together, and making special sounds at their task—
such as builders, soldiers, and sailors—are widely represented in modern speech.

Though mimicry, sometimes exact, but probably
more often a mere echo or suggestion of the sound to
be recalled, is responsible for some of the material of
Language, multitudes of words appear to have no such
origin. There are infinitely more words than sounds
in the world; and even things which have very distinct sounds have been named without any regard to
them. The inventors of the word *watch*, for instance,
did not call it *tick-tick* but *watch*, the idea being taken
from the *watchman* who walked about at night and
kept the time; and when the steam-engine appeared,
instead of taking the obvious sound-name *puff-puff*, it
was called *engine* (Lat. *ingenium*), to signify that it
was a work of *genius*. These modern words, however,
are the coinages of an intellectual age, and it was to
be expected that the inventors should look deeper
below the surface. How those words which have no
apparent association with sound were formed in early
times remains a mystery. With some the original
sound-association has probably been lost; in the case
of others, the association may have been so indirect as
to be now untraceable. The sounds available in savage life for word-making could never have been so
numerous as the things requiring names, and as civilization advanced the old words would be used in new
connections, while wholly new terms must have been

coined from time to time. Both these methods—the habit of generalizing unconsciously from single terms, and the trick of coining new words in a wholly conventional way—are still continually employed by savages as well as by children. Thus, to take an example of the first, Mr. John Moir, one of the earliest white men to settle in East Central Africa, was at once named by the natives *Mandala*, which means "a reflection in still water," because he wore on his eyes what looked to them a *still water* (spectacles). Afterwards they came to call not only Mr. Moir by that name, but spectacles, and finally—when it entered the country—glass itself. Examples of generalization among children abound in every nursery. A child is taken to the window by his nurse to see the moon. The easy monosyllable is caught up at once, and for some time the child applies it indiscriminately to anything bright or shining—the gas, the candle, the firelight are each "the moon." Mr. Romanes records a case where a child made a similar use of the word star —the gas, the candle, the firelight were each "a star." If the makers of Language proceeded on this principle, no wonder the philologist has riddles to read. How often must the savage children of the world have started off naming things from two such different points ? Mr. Romanes mentions a still more elaborate example which was furnished him by Mr. Darwin : " The child, who was just beginning to speak, called a duck 'quack,' and, by special association, it also called water 'quack.' By an appreciation of the resemblance of qualities, it next extended the term 'quack' to denote all birds and insects on the one hand, and all fluid substances on the other. Lastly, by a still more

delicate appreciation of resemblance, the child eventu-
ally called all coins ' quack,' because on the back of a
French sou it had once seen the representation of an
eagle. Hence, to the child, the sign ' quack,' from
having originally had a very specialized meaning, became more and more extended in its significance, until
it now seems to designate such apparently different
objects as ' fly,' ' wine,' and ' coin.' " [1]

The instructiveness of this, in showing the reason
why philology is often so helplessly at a loss in track-
ing far-strayed words to their original sense, is plain.
In the nature of the case, the onomatopoetic theory
can never be proved in more than a fraction of cases.
So cunning is the mind in associating ideas, so swift
in making new departures, that the clue to multitudes
of words must be obliterated by time, even if the first
forms and spellings of the words themselves remain
in their original integrity—which rarely happens—to
offer a feasible point to start the search from.

But it is far from necessary to assume that all
words should have had a rational ancestry. On the
contrary many words are probably deliberate artificial inventions. When not only every human being,
but every savage and every child has the ability as
well as the right to call anything it likes by any name
it chooses, it is vain in every case to seek for any general principle underlying the often arbitrary conjunc-
tions of letters and sounds which we call words.
Words cannot all at least be treated with the same
scientific regard as we would treat organic forms.
When dissected, in the nature of the case, they cannot
be expected to reveal specific structure such as one

[1] *Mental Evolution*, p. 283.

finds in a fern or a cray-fish. A fern or a cray-fish is the expression of an infinitely subtle and intricate adaptation, while a word may be a mere caprice. Perhaps, indeed, the greatest marvel about philology is that there should be a philology at all—that Languages should be so rich in association, so pregnant with the history and poetry of the past. Into the problem, therefore, of how the infinite variety of words in a Language was acquired it is unnecessary to enter at length. Once the idea had dawned of expressing meaning by sounds, the formation of words and even of Languages is a mere detail. We have probably all invented words. Almost every family of children invents words of its own, and cases are known where quite considerable Languages have been manufactured in the nursery. When boys play at brigands and pirates they invent pass-words and names, and from mere love of secrets and mysteries concoct vocabularies which no one can understand but themselves.

This simple fact indeed has been used with great plausibility to account for differences in dialect among different tribes, and even for the partial origin of new Languages. Thus the structure of the Indian languages has long puzzled philologists. Whitney informs us that as regards the material of expression, there is "irreconcilable diversity" among them. "There are a very considerable number of groups between whose significant signs exist no more apparent correspondences than between those of English, Hungarian, and Malay; none namely which may not be merely fortuitous." To account for these dialects a suggestion, as interesting as it is ingenious, has been

advanced by Dr. Hale. Imagine the case of a family
of Red Indians, father, mother, and half a dozen
children, in the vicissitudes of war, cut off from their
tribe. Suppose the father to be scalped and the
mother soon to die. The little ones left to themselves
in some lonely valley, living upon roots and herbs,
would converse for a time by using the few score
words they had heard from their parents. But as
they grew up they would require new words and
would therefore coin them. As they became a tribe
they would require more words, and so in time a Language might arise, all the words expressive of the
simpler relations—father, mother, tent, fire—being
common to other Indian Languages, but all the later
words purely arbitrary and necessarily a standing
puzzle to philology. The curious thing is that this
theory is borne out by some most interesting geographical facts. " If, under such circumstances, disease, or the casualties of a hunter's life should carry
off the parents, the survival of the children would, it
is evident, depend mainly upon the nature of the
climate and the ease with which food could be pro-
cured at all seasons of the year. In ancient Europe,
after the present climatal conditions were established,
it is doubtful if a family of children under ten years
of age could have lived through a single winter. We
are not, therefore, surprised to find that no more than
four or five linguistic stocks are represented in
Europe. Of North America, east of the Rocky Mountains and north of the tropics, the same may be said.
The climate and the scarcity of food in winter forbid
us to suppose that a brood of orphan children could
have survived, except possibly, by a fortunate chance,

in some favored spot on the shore of the Mexican Gulf, where shell-fish, berries, and edible roots are abundant and easy of access. But there is one region where Nature seems to offer herself as the willing nurse and bountiful stepmother of the feeble and unprotected. Of all countries on the globe, there is probably not one in which a little flock of very young children would find the means of sustaining existence more readily than in California. Its wonderful climate, mild and equable beyond example, is well known. Half the months are rainless. Snow and ice are almost strangers. There are fully two hundred cloudless days in every year. Roses bloom in the open air through all seasons. Berries of many sorts are indigenous and abundant. Large fruits and edible nuts on low and pendant boughs may be said in Milton's phrase to 'hang amiable.' Need we wonder that in such a mild and fruitful region, a great number of separate tribes were found speaking languages which careful investigation has classed in nineteen distinct linguistic stocks?" [1] Even more striking is the case of Oregon on the Californian border, which is also a favored and luxuriant land. The number of linguistic stocks in this narrow district is more than twice as large as in the whole of Europe.[2]

[1] Dr. Hale. Cf. Romanes, *Mental Evolution in Man*, p. 260.

[2] The construction of the mouth and lips has of course had something to do with differences in Languages, and even with the possibility of language in the case of Man. You must have your trumpet before you can get the sound of a trumpet. One reason why many animals have no speech is simply that they have not the mechanism which by any possibility could produce it. They might have a Language, but nothing at all like human Language. It is one of the significant notes in Evolution that

In such ways as these we may conceive of early Man building up the fabric of speech. In time his vocabulary would enlarge and become, so far as objects in the immediate environment were concerned, fairly complete. As Man gained more knowledge of the things around him, as he came into larger relations with his fellows, as life became more rich and complex, this accumulation of words would go on, each art as it was introduced creating new terms, each science pouring in contributions to the fund, until the materials of human speech became more and more complete. This process was never finished. The evolution of Language is still going on. No corroboration of the theory of the evolution of Language could be more perfect than the simple fact that it has gone on steadily down to the present hour and is going on now. Tens of thousands of words—no longer now onomatopoetic—have been evolved since Johnson com-

Man, almost alone among vertebrates, has a material body so far developed as to make it an available instrument for speech. There was almost certainly a time when this was to him a physical impossibility.

"The acquisition of articulate speech," says Prof. Macalaster, "became possible to man only when the alveolar arch and palatine area became shortened and widened, and when his tongue, by its accommodation to the modified mouth, became shorter and more horizontally flattened, and the higher refinements of pronunciation depend for their production upon the more extensive modifications in the same direction." Even for differences in dialect, as the same writer points out, there is a physical basis. "With the macrodont alveolar arch and the corresponding modified tongue, sibilation is a difficult feat to accomplish, and hence the sibilant sounds are practically unknown in all the Australian dialects."—British Association : Anthropological Section. Edinb., 1891.

piled his dictionary, and every year sees additions not only to technical terms but to the language of the people. The English Language is now being grown on two or three different kinds of soil, and the different fruits and flavors that result are intercharged and mixed, to enrich, or adulterate, the common English tongue. The mere fact that Language-making is a living art at the present hour, if not an argument against the theory that Language is a special gift, at least shows that Man has a special gift of making Language. If Man could manufacture words in any quantity, there was little reason why he should have been presented with them ready-made. The power to manufacture them is gift enough, and none the less a gift that we know some of the steps by which it was given, or at least through which it was exercised. But if the very words were given him as they stand, it is more than singular that so many of them should bear traces of another origin. Even Trench at this point succumbs to the theory of development, and his testimony is the more valuable that it is evidently so very much against the grain to admit it. He begins by stating apparently the opposite:—"The truer answer to the inquiry how language arose is this: God gave man language just as He gave him reason, and just because He gave him reason; for what is man's *word* but his *reason* coming forth that it may behold itself? They are indeed so essentially one and the same that the Greek language has one word for them both. He gave it to him, because he could not be man, that is, a social being, without it." Yet he is too profound a student of words to fail to qualify this, and had he failed to do so every page in his well-known book had judged him.

12

" Yet," he continues, "this must not be taken to affirm that man started at the first furnished with a full-formed vocabulary of words, and as it were with his first dictionary and first grammar ready-made to his hands. He did not thus begin the world with *names*, but *with the power of naming:* for man is not a mere speaking machine; God did not teach him words, as one of us teaches a parrot, from without; but gave him a capacity, and then evoked the capacity which he gave.[1]"

If the theory just given as to the formation of Language, or at least as to the possible formation of Language, be more than a fairy tale, there is another quarter in which corroboration of an important kind should lie. Hitherto we have examined as witnesses, the makers of words; it may be worth while for a moment to place in the witness-box the words themselves. A chemist has two methods of determining the composition of any body, analysis and synthesis. Having seen how words may be built up, it remains for us to see whether on analysis they bear trace of having been built up in the way, and from the elements, suggested. Comparative Philology has now made an actual investigation into the words and structure of all known Languages, and the information sought by the evolutionist lies ready-made to his hand. So far as controversy might be expected to arise here on the theory of development itself, there is none. For the first fact to interest us in this new region is that every student of Language seems to have been compelled to give in his adherence to the general theory of Evolution. All agree with Renan

[1] Archbishop Trench, *The Study of Words*, pp. 14, 15.

that " Sans doubte les langues, comme tout ce qui est organisé, sont sujettes à la loi du dévelopment graduel." And even Max Müller, the least thorough-going from an evolutionary point of view of all philologists, asserts that " no student of the science of Language can be anything but an evolutionist, for, wherever he looks, he sees nothing but evolution going on all around him."

The outstanding discovery of the dissector of words is that, vast and complex as Languages appear, they are really composed of few and simple elements. Take the word "evolutionary." The termination " ary " is a late addition added to this and to thousands of other words for a special purpose; the same applies to the syllable "tion." The first letter *e* distinguishes evolution from convolution, revolution, involution, and is also a later growth. None of these extra syllables is of first importance; by themselves they have almost no meaning. The part which will *not* disappear or melt away into mere grammar, on which the stress of the sense hangs, is the syllable " vol " or " volv," and, so far as the English language is concerned, it is to be looked upon as the root. By running it to earth in older languages its source is found in a still more radical word, and therefore it must next be blotted out of the list of primitive words. By patient comparison of all other words with all other words, of Languages with Languages, and apparent roots with apparent roots, the supposed primitive roots of Language have been found. Just as all the multifarious objects in the material world—water, air, earth, flesh, bone, wood, iron, paper, cloth—are resolvable by the chemist into some sixty-eight elements, so all the words in each of the three or four great groups of Language yield on the last analysis only a few hundred original roots. That still further analysis may break down some or many of these is not impossible. But the facts as they stand are all significant. The further we go back into the past the Languages become thinner and thinner, the words fewer and fewer, the grammar poorer and poorer. Of the thousand known Languages it has been found possible to reduce all to three or four—probably three—great families; and each of these in turn is capable of almost unlimited philological pruning. In analyzing the Sanskrit language, Professor Max Müller reduces its whole vocabulary to 121 roots—the 121 "original concepts." "These 121 concepts constitute the stock-in-trade with which I maintain that every thought that has ever passed through the mind of India, so far as known to us in its literature, has been expressed. It would have been easy to reduce that number still further, for there are several among them which could be ranged together under more general concepts. But I leave this further reduction to others, being satisfied as a first attempt with having shown how small a number of seeds may produce, and has produced, the enormous intellectual vegetation that has covered the soil of India from the most distant antiquity to the present day." [1]

That a "first attempt" should have succeeded in reducing this vast family of Languages to 121 words is significant. The exhumation by philology of this early cluster reminds one of the discovery of the seg-

---

[1] *Science of Thought*, p. 549.

mented ovum in embryology. Such clusters appear at an early stage in the history of all developments. The processes which precede this stage are of the utmost subtlety, but in embryology they have yielded to the latter analysis of the microscope. So it may be one day with the natural history of Language. We may never, for obvious reasons, get back to the actual beginning, but we may get nearer. When the embryologist reached his cluster of cells in the segmented ovum, he did not believe he had found the dawn of life. What further the philologist may find remains a mystery. Where these 121 words came from may never be known. But the development from that point sufficiently shows that words, like everything else, have followed the universal law, and that Languages, starting from small beginnings, have grown in volume, intricacy, and richness, as time rolled on. "All philologists," says Romanes, "will now agree with Geiger—'Language diminishes the further we look back, in such a way that we cannot forbear concluding it must once have had no existence at all.'"

The history of progress for a long time henceforth is the history of the progress of Language and the increase in intelligence which necessarily went along with it. From being able to say what he knew, Man went on to write what he knew. The Evolution of writing went through the same general stages as the Evolution of Speech. First there was the onomatopoetic writing—as it were, the growl-writing—the ideograph, the imitation of an actual object. This is the form we find fossil in the Egyptian hieroglyphic. For a man a man was drawn, for a camel a camel, for a hut a hut. Then intonation was added—accents, that

is, for extra meaning or extra emphasis. Then to
save time the objects were drawn in shorthand—a
couple of dashes for the limbs and one across, as in
the Chinese for man; a square in the same language
for a field; two strokes at an obtuse angle, suggesting
the roof, for a house. To express further qualities,
these abbreviated pictures were next compounded in
ingenious ways. A man and a field together conveyed
the idea of wealth, and because a man with a field was
rich, he was supposed to be happy, and the same combination stood, and stands to this day, for content-
ment. When a roof is drawn and a woman beneath
it—or the strokes which represent a roof and a woman
—we have the idea of a woman at home, a woman at
peace, and hence the symbol comes to stand for quiet-
ness and rest. Chinese writing is picture-writing,
with the pictures degenerated into dashes—a lingual
form of the modern impressionism.

When writing was fully evolved, this height was
only the starting-point for some new development.
Every summit in Evolution is the base of some
grander peak. Speech, whether by writing or by
spoken word, is too crude and slow to keep pace with
the needs of the now swiftly ascending mind. Man's
larger life demands a further specialization of this
power. He learned to speak at first because he
could not convey his thoughts to his wife at the other
side of the wood. It was Space that made him speak.
He now learns to speak better because he cannot convey his thoughts to the other end of the world. This
new distance-language began again at the beginning,
just as all Language does, by employing signs. Man
invented the telegraph—a little needle which makes

signs to some one at the other side of the world. The telegraph is a gesture-language, and is therefore only a primitive stage. Man found this out and from signs went on to sounds—he invented the telephone. By all the traditions of Evolution this marvellous instrument ought to be, and is even now on the verge of being, the vehicle of the distance-language of the future.

Is this the end? It is by no means likely. The mind is feeling about already for more perfect forms of human intercourse than telegraphed or telephoned words. As there was a stage in the ascent of Man at which the body was laid aside as a finished product, and made to give way to Mind, there may be a stage in the Evolution of Mind when its material achievements—its body—shall be laid aside and give place to a higher form of Mind. Telepathy has already become a word, not a word for thought-reading or muscle-reading, but a scientific word. It means "the ability of one mind to impress, or to be impressed by another mind otherwise than through the recognized channels of sense."[1] By men of science, adepts in mental analysis, aware of all sources of error, armed against fraud, this subject is now being made the theme of exhaustive observation. It is too soon to pronounce. Practically we are in the dark. But there are those in this fascinating and mysterious region who tell us that the possibilities of a more intimate fellowship of man with man, and soul with soul, are not to be looked upon as settled by our present views of matter or of mind. However little we know of it, however remote we are from it, whether it ever be realized or not, telepathy is theoretically the

[1] *Phantasms of the Living*, p. 6.

next stage in the Evolution of Language. As we have
seen, the introduction of speech into the world was
delayed, not because the possibilities of it were not in
Nature, but because the instrument was not quite ·
ready. Then the instrument came, and Man spoke.
The development of the organ and the development of
the function went on together, arrived together, were
perfected together. What delayed the gesture-language of the telegraph was not that electricity was
not in Nature, but the want of the instrument.
When that came, the gesture-language came, and both
were perfected together. What delayed the telephone
was not that its principle was not in Nature, but that
the instrument was not ready. What now delays its
absolute victory of space is not that space cannot be
bridged, but that it is not ready. May it not be that
that which delays the power to transport and drive
one's thought as thought to whatever spot one wills,
is not the fact that the possibility is withheld by
Nature, but that the hour is not quite come—that the
instrument is not yet fully ripe? Are there no signs,
is the feeling after it no sign, are there not even now
some facts, to warrant us in treating it, after all that
Evolution has given us, as a still possible gift to the
human race? What strikes one most in running the
eye up this graduated ascent is that the movement is
in the direction of what one can only call spirituality.
From the growl of a lion we have passed to the
whisper of a soul; from the motive fear, to the motive
sympathy; from the icy physical barriers of space, to
a nearness closer than breathing; from the torturing
slowness of time to time's obliteration. If Evolution
reveals anything, if science itself proves anything, it

is that Man is a spiritual being and that the direction
of his long career is towards an ever larger, richer,
and more exalted life. On the final problem of Man's
being the voice of science is supposed to be dumb.
But this gradual perfecting of instruments, and, as
each arrives, the further revelation of what lies behind in Nature, this gradual refining of the mind, this
increasing triumph over matter, this deeper knowledge, this efflorescence of the soul, are facts which even
Science must reckon with. Perhaps, after all, Victor
Hugo is right: "I am the tadpole of an archangel."

Before closing this outline two of the many omitted points may be briefly referred to. In thinking of
Language as a "discovery," it is not necessary to assume that that discovery involved the pre-existence
of very high mental powers. These were probably
developed *pari passu* with Speech, but did not necessarily ante-date it to such a degree as to make the
preceding argument a *petitio principii*. Obviously the
discovery of Language could not in the first instance
have been responsible for the Evolution of Mind, since
Man must already have had Mind enough to discover
it. But this does not necessarily imply any very high
grade of intellect—very high, that is to say, as compared with other contemporary animals—for it is possible that a comparatively slight rise in intelligence
might have led to the initial step from which all the
others might follow in rapid succession. An illustration, suggested by a remark of Cope's, may help to
make plain how a very slight cause may initiate
changes of an almost radical order and on the most
gigantic scale.

In part of the Arctic regions at this moment there is no such thing as liquid. Matter is only known there in the solid form. The temperature may be thirty-one degrees below zero, or thirty-one degrees above zero without making the slightest difference; there can be nothing there but ice, glacier, and those crystals of ice which we call snow. But suppose the temperature rose two degrees, the difference would be indescribable. While no change for sixty degrees below that point made the least difference, the almost inappreciable addition of two degrees changes the country into a world of water. The glaciers, under the new conditions, retreat into the mountains, the vesture of ice drops into the sea, a garment of greenness clothes the land. So, in the animal world, a very small rise beyond the animal maximum may open the door for a revolution. With a brain of so many cubic inches, and so many pounds of brain matter, we have animal intelligence. Everything below that limit is animal, and the number of inches or pounds below it makes no difference. But pass to a brain not a few but many pounds heavier, many cubic inches larger, and very much more convoluted, and it is conceivable that in passing from the lower to the higher figures some such change might occur as that which differentiates solid from liquid in the case of water. What the chemist calls a "critical point" might thus be passed, and from a condition associated with certain properties—though in the brain we must speak of accompaniments rather than properties—a condition associated with certain other properties might be the result. Thus, as Cope says, " some Rubicon has been crossed, some flood-gate has been opened, which marks one of Nature's great transitions, such as have been called 'expression-points' of progress." A slight rise in intelligence might lead to the first acquisition of Speech, and from this point the rise might be at once exceedingly swift and in directions wholly new. The illustration is not to be taken for more than it seeks to illustrate—which is not the method of transition as to qualitative detail, but simply the fact that an apparently slight change may have startling and indefinite results.

The last difficulty is this. If the connection between Mind and Language is so vital, why do not Birds, many of which apparently speak, emulate Man in mental power? If his speech is largely responsible for his intelligence, why have not Birds—the parrot, for instance—attained the same intelligence? Several answers might be suggested to the question, and several kinds of answers—biological, physiological, philological, and psychological. But the real answer is the general one, that to make animals human required a conspiracy of circumstances which neither Birds nor any other animal fell heir to. It was one chance in a million that the multitude of co-operating conditions which pushed Man onward were fulfilled; and though it may never be known what these conditions were, it was doubtless from the failure on the one hand to meet one or more of them, and on the other from the success with which openings in other directions were pursued by competing species, that Man was left alone during the later æons of his ascent.

The progenitors of Birds and the progenitors of Man at a very remote period were probably one. But at a certain point they parted company and diverged hopelessly and forever. The Birds took one road, the Vertebrates another ; the Vertebrates kept to the ground, the Birds took to the air. The consequences of this expedient in the case of the Birds were fatal. They forever forfeited the possibility of becoming human. For observe the cost to them of the aerial mode of life. The wing was made *at the expense of the hand.* With this consummate organ buried in feathers, the use which the higher Vertebrates made of it was denied them. Birds have the bones for a hand, could have had a hand, but they waived their right to it. When it is considered how much Man owes to the hand it may be conceived how much they have lost by the want of it. Had Man not been a "tool-using animal," he had probably never become a man ; the Bird, partly because it placed itself out of the running here, has never been anything but a Bird. To one organism only was it given to keep on the path of progress from the beginning to the end, and so fulfil without deviation or relapse the final purpose of Evolution.

CHAPTER VI.

THE STRUGGLE FOR LIFE.

Matthew Arnold, in a well-remembered line, describes a bird in Kensington Gardens " deep in its unknown day's employ." But, peace to the poet, its employ is all too certain. Its day is spent in struggling to get a living; and a very hard day it is. It awoke at daybreak and set out to catch its morning meal; but another bird was awake before it, and it lost its chance. With fifty other breakfastless birds, it had to bide its time, to scour the country; to prospect the trees, the grass, the ground; to lie in ambush; to attack and be defeated; to hope and be forestalled. At every meal the same programme is gone through, and every day. As the seasons change the pressure becomes more keen. Its supplies are exhausted, and it has to take wing for hundreds and thousands of miles to find new hunting-ground. This is how birds live, and this is how birds are made. They are the children of Struggle. Beak and limb, claw and wing, shape, strength, all down to the last detail, are the expressions of their mode of life.

This is how the early savage lived, and this is how he was made. The first practical problem in the

189

Ascent of Man was to get him started on his upward path. It was not enough for Nature to equip him with a body, and plant his foot on the lowest rung of the ladder. She must introduce into her economy some great principle which should secure, not for him alone but for every living thing, that they should work upward toward the top. The inertia of things is such that without compulsion they will never move. And so admirably has this compulsion been applied that its forces are hidden in the very nature of life itself—the very act of living contains within it the principles of progress. An animal cannot *be* without *becoming*.

The first great principle into the hands of which this mighty charge was given is the Struggle for Life. It is one of the chief keys for unlocking the mystery of Man's Ascent, and so important in all development that Mr. Darwin assigns it the supreme rank among the factors in Evolution. " Unless," he says, " it be thoroughly engrained in the mind, the whole economy of Nature, with every fact on distribution, rarity, abundance, extinction, and variation, will be dimly seen or quite misunderstood." How, under the pressures of this great necessity to work for a living, the Ascent of Man has gone on, we have now to inquire. Though not to the extent that is usually supposed, yet in part under this stimulus, he has slowly emerged from the brute-existence, and, entering a path where the possibilities of development are infinite, has been pushed on from stage to stage, without premeditation, or design, or thought on his part, until he arrived at that further height where, to the unconscious compulsions of a lower environment, there were added those high incitements of conscious ideals which completed the work of creating him a Man.

Start with a comparatively unevolved savage, and see what the Struggle for Life will do for him. When we meet him first he is sitting, we shall suppose, in the sun. Let us also suppose—and it requires no imagination to suppose it—that he has no wish to do anything else than sit in the sun, and that he is perfectly contented, and perfectly happy. Nature around him, visible and invisible, is as still as he is, as inert apparently, as unconcerned. Neither molests the other; they have no connection with each other. Yet it is not so. That savage is the victim of a conspiracy. Nature has designs upon him, wants to do something to him. That something is to move him. Why does it wish to move him? Because movement is work, and work is exercise, and exercise may mean a further evolution of the part of him that is exercised. How does it set about moving him? By moving itself. Everything else being in motion, it is impossible for him to resist. The sun moves away to the west and he must move or freeze with cold. As the sun continues to move, twilight falls and wild animals move from their lairs and he must move or be eaten. The food he ate in the morning has dissolved and moved away to nourish the cells of his body, and more food must soon be moved to take its place or he must starve. So he starts up, he works, he seeks food, shelter, safety; and those movements make marks in his body, brace muscles, stimulate nerves, quicken intelligence, create habits, and he becomes more able and more willing to repeat these movements and so becomes a stronger and a higher man. Multiply these

movements and you multiply him. Make him do
things he has never done before, and he will become
what he never was before. Let the earth move round
in its orbit till the sun is far away and the winter
snows begin to fall. He must either move away, and
move away very fast, to find the sun again ; or he must
chase, and also very fast, some thick-furred animal,
and kill it, and clothe himself with its skin. Thus
from a man he has become a hunter, a different kind
of a man, a further man. He did not wish to become
a hunter; he *had* to become a hunter. All that he
wished was to sit in the sun and be let alone, and but
for a Nature around him which would not rest, or let
him alone, he would have sat on there till he died.
The universe has to be so ordered that that which Man
would not have done alone he should be compelled to
do. In other words it was necessary to introduce into
Nature, and into Human Nature, some such principle
as the Struggle for Life. For the first law of Evolution is simply the first law of motion. " Every body
continues in a state of rest, or of uniform motion in a
straight line, unless it is compelled by impressed forces
to change that state." Nature supplied that savage
with the impressed forces, with something which he
was compelled to respond to. Without that, he would
have continued forever as he was.

Apart from the initial appetite, Hunger, the stimulus of Environment—that which necessitates Man to
struggle for life—is twofold. The first is inorganic
nature, including heat and cold, climate and weather,
earth, air, water—the material world. The second is
the world of life, comprehending all plants and animals, and especially those animals against whom prim-

itive Man has always to struggle most—other primitive Men. All that Man is, all the arts of life, all the gifts of civilization, all the happiness and joy and progress of the world, owe much of their existence to that double war.

Follow it a little further. Go back to a time when Man was just emerging from the purely animal state, when he was in the condition described by Mr. Darwin, "a tailed quadruped probably arboreal in its habits," and when in his glimmering consciousness mind was feeling about for its first uses in snatching some novel success in the Struggle for Life. This hypothetical creature, so far as bodily structure was concerned, was presumably not very vigorous. Had he been more vigorous he might never have evolved at all; as it was, he fled for refuge not to his body but to a stratagem of the Mind. When threatened by a comrade, or pressed by an alien-species, he called in a simple foreign aid to help him in the Struggle—the branch of a tree. Whether the discovery was an accident; whether the idea was caught from the falling of a bough, or a blow from a branch waving in the wind, is of no consequence. This broken branch became the first *weapon*. It was the father of all *clubs*. The day this discovery was made, the Struggle for Life took a new departure. Hitherto animals fought with some specialized part of their own bodies—tooth, limb, claw. Now they took possession of the armory of material Nature.

This invention of the club was soon followed by another change. To use a club effectively, or to keep a good look-out for enemies or for food, a man must stand erect. This alters the centre of gravity of the

13

body, and as the act becomes a habit, subsidiary changes slowly take place in other parts. In time the erect position becomes confirmed. Man owes what Burns calls his " heaven-erected face " to the Struggle for Life. How recent this change is, how new the attitude still is to him, is seen from the simple fact that even yet he has not attained the power of retaining the erect position long. Most men sit down when they can, and so unnatural is the standing position, so unstable the equilibrium, that when slightly sick or faint, Man cannot stand at all.

Possibly both the erect position and the Club had another origin, but the detail is immaterial. This " hairy-tailed quadruped, arboreal in its habits," must sometimes have wandered or been driven into places where trees were few and far between. It is conceivable that an animal, accustomed to get along mainly by grasping something, should have picked up a branch and held it in its hand, partly to use as a crutch, partly as a weapon, and partly to raise itself from the ground in order to keep a better look-out in crossing treeless spaces. An Orang-outang may now be seen in the Zoological gardens in Java, which promenades about its bower continually with the help of a stick, and seems to prefer the erect position so long as the stick or any support is at hand.

The next stage after the invention of anything is to improve upon it, or to make a further use of it. Both these things now happened. One day the stick, wrenched rapidly from the tree, happened to be left with a jagged end. The properties of the *point* were discovered. Now there were two classes of weapons

in the world—the blunt stick and the pointed stick—
that is to say, the Club and the Spear.

In using these weapons at first, neither probably
was allowed to leave the hand. But already their
owners had learned to hurl down branches from the
tree-tops, and bombard their enemies with nuts and
fruits. Hence they came to throw their clubs and
spears, and so *missiles* were introduced. Under this
new use, the primitive weapons themselves received
a further specialization. From the heavy bludgeon
would arise on the one hand the shaped war-club, and
on the other the short throwing club, or waddy. The
spear would pass into the throwing assegai, or the
ponderous weapon such as the South Sea Islanders
use to-day. From the natural point of a torn branch
to the sharpening of a point deliberately is the next
improvement. From rubbing the point against the
sharp edge of a large stone, to picking up a sharp-
edged small stone and using it as a knife, is but a
step. So, by the mere necessities of the Struggle for
Life, development went on. Man became a tool-using
animal, and the foundations of the Arts were laid.
Next, the man who threw his missile furthest, had the
best chance in the Struggle for Life. To throw to
still greater distances, and with greater precision, he
sought out mechanical aids—the bow, the boomerang,
the throwing-stick, and the sling. Then instead of
using his own strength he borrowed strength from
nature, mixed different kinds of dust together and
invented gunpowder. All our modern weapons of
precision, from the rifle to the long range gun, are
evolutions from the missiles of the savage. These
suggestions are not mere fancies ; in savage tribes

existing in the world to-day these different stages in Evolution may still be seen.

After weapons of offence came weapons of defence. At first the fighting savage sheltered himself at the back of a tree. Then when he wished to pass to another tree he tore off part of the bark, took it with him, and made the first *shield.* Where the trees were without suitable bark, he would plait his shield from canes, grasses, and the midribs of the leaves, or construct them from frameworks of wood and skins. In times of peace these hollow shields, lying idly about the huts, would find new uses—baskets, cradles, and, in an evolved form, coracles or boats. In leisure hours also, new virtues discovered themselves in the earlier implements of war and of the chase. The twang of his bow suggested memories that were pleasant to his ear; he kept on twanging it, and so made *music.* Because two bows twanged better than one, he twanged two bows; then he made himself a two-stringed bow from the first, and ended with a "ten-stringed instrument." By and bye came the harp; later, the violin. The whistling of the wind in a hollow reed prepared the way for the flute; a conchshell, broken at the helix, gave him the trumpet. Two flints struck together yielded fire.

Trifling, almost puerile, as these beginnings look to us now, remember they were once the serious realities of life. The club and spear of the savage are toys to us to-day; but we forget that the rude shafts of wood which adorn our halls were all the world to early Man and represented the highest expression and daily instrument of his evolution. These primitive weapons are the pathetic expression of the world's first Struggle. As the earliest contribution of mankind to solve its still fundamental difficulty—the problem of Nutrition—they are of enduring interest to the human race. So far from being, as one might suppose, mere implements of destruction, they are implements of self-preservation; they entered the world not from hate of Man but for love of life. Why was the spear invented, and the sling, and the bow? In the first instance because Man needed the bird and the deer for food. Why from implements of the chase did they change into implements of war? Because other men wanted the bird and the deer, and the first possessor, as populations multiplied, must protect his food-supply. The parent of all industries is Hunger: the creator of civilization in its earlier forms is the Struggle for Life.

By hollowing a pit in the ground, planting his spear, or a pointed stake, upright in the centre, and covering the mouth with boughs, Man could trap even the largest game. When the climate became cold, he stripped off the skin and became the possessor of clothes. With a stone for a hammer, he broke open molluscs on the shore, or speared or trapped the fish in the shoals. Digging for roots with his pointed stick in time suggested agriculture. From imitating the way wild fruits and grains were sown by Nature he became a gardener and grew crops. To possess a crop means to possess an estate, and to possess an estate is to give up wandering and begin that more settled life in which all the arts of industry must increase. Catching the young of wild animals and keeping them, first as playthings, then for supplies of meat or milk, or, in the case of the dog, for helping in

the chase, he perceived the value of domestic animals.
So Man slowly passed from the animal to the savage,
so his mind was tamed, and strengthened, and bright-
ened, and heightened; so the sense of power grew
strong, and so *virtus*, which is to say virtue, was
born.

In struggling with Nature, early Man not only
found material satisfactions: he found *himself.* It
was this that made him, body, mind, character, and
disposition; and it was this largely that gave to the
world different kinds of men, different kinds of bodies,
minds, characters, and dispositions. The first moral
and intellectual diversifiers of men are to be sought for
in geography and geology—in the factors which determine the circumstances in which men severally conduct their Struggle for Life. If the land had been all
the same, the Struggle for Life had been all the same,
and if the Struggle for Life had been all the same, life
itself had been all the same. But to no two sets of
men is the world ever quite the same. The theatre
of struggle varies with every degree of latitude, with
every change of altitude, with every variation of soil.
In most countries three separate regions are found—
a maritime region, an agricultural region, a pastoral
region. In the first, the belt along the shore, the
people are fishermen; in the second, the lowlands
and alluvial plains, the people are farmers; in the
third, the highlands and plateaux, they are shepherds.
As men are nothing but expressions of their environments, as the kind of life depends on how men
get their living, each set of men becomes changed
in different ways. The fisherman's life is a precarious life; he becomes hardy, resolute, self-re-

liant. The farmer's life is a settled life; he becomes tame, he loves home, he feeds on grains and fruits which take the heat out of his blood and make him domestic and quiet. The shepherd is a wanderer; he is much alone; the monotonies of grass make him dull and moody; the mountains awe him: the protector of his flock, he is a man of war. So arise types of men, types of industries; and by and bye, by exogamous marriage, blends of these types, and further blends of infinite variety. "It is so ordered by Nature, that by so striving to live they develop their physical structure; they obtain faint glimmerings of reason; they think and deliberate; they become Man. In the same way, the primeval men have no other object than to keep the clan alive. It is so ordered by Nature that in striving to preserve the existence of the clan, they not only acquire the arts of agriculture, domestication, and navigation: they not only discover fire, and its uses in cooking, in war, and in metallurgy; they not only detect the hidden properties of plants, and apply them to save their own lives from disease, and to destroy their enemies in battle; they not only learn to manipulate Nature and to distribute water by machinery; but they also, by means of the life-long battle, are developed into moral beings." [1] Nature being "everything that is," and Man being in every direction immersed in it and dependent on it, can never escape its continuous discipline. Some environment there must always be; and some change of environment, no matter how minute, there must always be; and some change, no matter how imperceptible, must be always wrought in him.

[1] Winwood Reade, *Martyrdom of Man*, p. 464.

We now see, perhaps, more clearly why Evolution at the dawn of life entered into league with so strange an ally as *Want.* The Evolution of Mankind was too great a thing to entrust to any uncertain hand. The advantage of attaching human progress to the Struggle for Life is that you can always depend upon it. Hunger never fails. All other human appetites have their periods of activity and stagnation ; passions wax and wane ; emotions are casual and capricious. But the continuous discharge of the function of Nutrition is interrupted only by the final interruption—Death. Death means, in fact, little more than an interference with the function of Nutrition ; it means that the Struggle for Life having broken down, there can be no more life, no further evolution. Hence, it has been ordained that Life and Struggle, Health and Struggle, Growth and Struggle, Progress and Struggle, shall be linked together ; that whatever the chances of misdirection, the apparent losses, the mysterious accompaniments of strife and pain, the Ascent of Man should be bound up with living. When it is remembered that, at a later day, Morality and Struggle, and even Religion and Struggle, are bound so closely that it is impossible to conceive of them apart, the tremendous value of this principle and the necessity for providing it with indestructible foundations, will be perceived.

This association of the Struggle for Life with the physiological function of Nutrition must be continually borne in mind. For the essential nature of the principle has been greatly obscured by the very name which Mr. Darwin gave to it. Probably no other was possible ; but the effect has been that men have emphasized the almost ethical substantive "Struggle" and ignored the biological term "Life." A secondary implication of the process has thus been elevated into the prime one; and this, exaggerated by the imagination, has led to Nature being conceived of as a vast murderous machine for the annihilation of the majority and the survival of the few. But the Struggle for Life, in the first instance, is simply living itself; at the best, it is living under a healthily normal maximum of pressure; at the worst, under an abnormal maximum. As we have seen, initially, it is but another name for the discharge of the supreme physiological function of Nutrition. If life is to go on at all, this function must be discharged, and continuously discharged. The primary characteristic of protoplasm, the physical basis of all life, is Hunger, and this has dictated the first law of being—"Thou shalt eat." What distinguishes scientifically the organic from the inorganic, the animal from the stone? That the animal eats, the stone does not. Almost all achievement in the early history of the living world has been due to Hunger. For millenniums nearly the whole task of Evolution was to perfect the means of satisfying it, and in so doing to perfect life itself. The lowest forms of life are little more than animated stomachs, and in higher groups the nutritive system is the first to be developed, the first to function, and the last to cease its work. Almost wholly, indeed, in the earlier vicissitude of the race, and largely in the more ordered course of later times, Hunger rules the life and work and destiny of men; and so profoundly does this mysterious deity still dominate the round of even the highest life that the noblest occupations which engage the human mind must be interrupted two or three times a day to do it homage.

Whatever Man came ultimately to wish and to achieve for himself, it was essential at first that such arrangements should be made for him. The machinery for his development had not only to be put into Nature, but he had to be placed in the machine and held there, and brought back there as often as he tried to evade it. To say that man evolved himself, nevertheless, is as absurd as to say that a newspaper prints itself. To say even that the machinery evolved him is as preposterous as to say of a poem that the printing-press made it. The ultimate problem is, Who made the machine? and Who thought the poem that was to be printed?

If you say that you do not unreservedly approve of the machine, that it lacerates as well as binds, the difficulty is more real. But it is a principle in the study of history to suspend judgment both of the meaning and of the value of a policy until the chain of sequences it sets in motion should be worked out to its last fulfilment. When the full tale of the Struggle for Life is told, when the record of its victories is closed, when the balance of its gains and losses has been struck, and especially when it is proved that there actually have been losses, it will be time to pass judgment on its moral value. Of course this principle cuts both ways; it warns off a favora le as well as an unfavorable verdict on the beneficence of the system of things. But Evolution is a study in history, and its results are largely known. And it would be affectation to deny that on the whole these results are good, and appear the worthier the more we penetrate into their inner meaning. Men forget when they denounce the Struggle for Life, that it is to be judged not only on the ground of sentiment but of reason, that not its local or surface effects only, but its permanent influence on the order of the world, must be taken into account.

Even on the lower ranges of Nature the unfavorable implications of the Struggle for Life have probably been exaggerated. While it is essential to an understanding of the course of evolution, to retain in the imagination a vivid sense of the Struggle itself, we must beware of over-coloring the representation, or flooding it with accompaniments of emotion borrowed from our own sensations. The word Struggle at all in this connection is little more than a metaphor. When it is said that an animal struggles, all that is really meant is that it lives. An animal, that is to say, does not, in addition to all its other activities, have to employ a vast number of special activities, to the exercise of which the term Struggle is to be applied. It is Life itself which is the Struggle: and the whole Life, and the whole of the activities and powers which make up life are involved in it. To speak of Struggle in the sense of some special and separate struggle, to conceive of battle, or even a series of battles, is misleading, where all is struggle and where all is battle. Especially must we beware of reading into it our personal ideas with regard to accompaniments of pain. The probabilities are that the Struggle for Life in the lower creation is, to say the least, less painful than it looks. Whether we regard the dulness of the states of consciousness among lower animals, or the fact that the condition

of danger must become habitual, or that death when
it comes is sudden, and unaccompanied by that anticipation which gives it its chief dread to Man, we
must assume that whatever the Struggle for Life
subjectively means to the lower animals, it can never
approach in terror what it means to us. And as to
putting any moral content into it, until a late stage
in the world's development, that is not to be thought
of. Judged of even by later standards there is much
to relieve one's first unfavorable impression. With
exceptions, the fight is a fair fight. As a rule there is
no hate in it, but only Hunger. It is seldom prolonged, and seldom wanton. As to the manner of
death, it is generally sudden. As to the fact of
death, all animals must die. As to the meaning of an
existence prematurely closed, it is better to be to be
eaten than not to be at all. And, as to the last
result, it is better to be eaten out of the world
and, dying, help another to live, than pollute the
world by lingering decay. The most, after all, that
can be done with life is to give it to others. Till
Nature taught her creatures of their own free will to
offer the sacrifice, is it strange that she took it by
force?

There are those indeed who frown upon Science
for predicating a Struggle for Life in Nature at all,
lest the facts should impugn the beneficence of the
universe. But Science did not invent the Struggle
for Life. It is there. What Science has really done
is to show not only its meaning but its great moral
purpose. There are others, again, like Mill, who, seeing the facts, but not seeing that moral purpose,
impugn natural theology for still believing in the

beneficence of that purpose. Neither attitude, probably, is quite worthy of the names with which these conclusions are associated. Much more reasonable are the verdicts of the two men who are first responsible for bringing the facts before the world, Mr. Alfred Russel Wallace and Mr. Darwin. " When we reflect," says Mr. Darwin, " on this struggle, we may console ourselves with the full belief that the war of nature is not incessant, that no fear is felt, that death is generally prompt, and that the vigorous, the healthy, and the happy survive and multiply." And in much stronger language Mr. Wallace : " On the whole, the popular idea of the struggle for existence entailing misery and pain on the animal world is the very reverse of the truth. What it really brings about is the maximum of life and of the enjoyment of life, with the minimum of suffering and pain. Given the necessity of death and reproduction, and without these there could have been no progressive development of the organic world—and it is difficult even to imagine a system by which a greater balance of happiness could have been secured." [1]

We may safely leave Nature here to look after her own ethic. That a price, a price in pain, and assuredly sometimes a very terrible price, has been paid for the evolution of the world, after all is said, is certain. There may be difference of opinion as to the amount of this price, but on one point there can be no dispute—that even at the highest estimate the thing which was bought with it was none too dear. For that thing was nothing less than the present progress of the world. The Struggle for Life has been a vic-

[1] *Darwinism,* pp. 30–40.

torious struggle; it has succeeded in its stupendous task; and there is nothing of order or beauty or perfection in living Nature that does not owe something to its having been carried on. The first duty of those who demur to the cost of progress is to make sure that they comprehend in all its richness the infinity of the gift this sacrifice has purchased for humanity. The end of the Struggle for Life is not battle; it is not even victory, it is evolution. The result is not wounds, it is health. Nature is a vast and complicated system of devices to keep things changing, adjusting, and, as it seems, progressing. The Struggle for Life is a species of necessitated aspiration, the *vis a tergo* which keeps living things in motion. It does not follow, of course, that that motion should be upward; that is dependent on other considerations. But the point to mark is that without the struggle for food and the pressure of want, without the conflict with foes and the challenge of climate, the world would be left to stagnation. Change, adventure, temptation, vicissitude even to the verge of calamity, these are the life of the world.

There is another side to this principle from which its higher significance becomes still more apparent. It follows from the Struggle for Life that those animals which struggle most successfully will prosper, while the less successful will disappear—hence the well-known principle of Natural Selection or the Survival of the Fittest. Waiving the discussion of this law in general, and the varying meanings which "fitness" assumes as we rise in the scale of being, observe the rôle it plays in Nature. The object of the Survival of the Fittest is to produce fitness. And it does so both negatively and positively. In the first place it produces fitness by killing off the unfit. Without the rigorous weeding out of the imperfect the progress of the world had not been possible. If fit and unfit indiscriminately had been allowed to live and reproduce their kind, every improvement which any individual might acquire would be degraded to the common level in the course of a few generations. Progress can only start by one or two individuals shooting ahead of their species; and their life-gain can only be conserved by their being shut off from their species—or by their species being shut off from them. Unless shut off from their species their acquisition will either be neutralized in the course of time by the swamping effect of inter-breeding with the common herd, or so diluted as to involve no real advance. The only chance for Evolution, then, is either to carry off these improved editions into "physiological isolation," or to remove the unimproved editions by wholesale death. The first of these alternatives is only occasionally possible; the second always. Hence the death of the unevolved, or of the unadapted in reference to some new and higher relation with environment, is essential to the perpetuation of a useful variation. Although Natural Selection by no means invariably works in the direction of progress,—in parasites it has consummated almost utter degeneration,—no progress can take place without it. It is only when one considers the working of the Struggle for Life on the large scale, and realizes its necessity to the Evolution of the world as a whole, that one can even begin to discuss its ethical or teleological meanings. To make a fit world, the unfit at every stage must be made to disappear; and

if any self-acting law can bring this about, though its
bearing upon this or that individual case may seem
unjust, its necessity for the world as a whole is vindicated. If more of any given species are born into the
world than can possibly find food, and if a given number must die, that number must be singled out upon
some principle ; and we cannot quarrel with the
principle in Physical Nature which condemns to death
the worst. By placing the death-penalty upon the
slightest short-coming, Natural Selection so discour-
ages imperfection as practically to eliminate it from
the world. The fact that any given animal is alive
at all is almost a token of its perfectness. Nothing
living can be wholly a failure. For the moment that
it fails, it ceases to live. Something more fit, were it
even by a hairbreadth, secures it place ; so that all
existing lives must, with reference to their environment, be the best possible lives. Natural Selection is
the means employed in Nature to bring about perfect
health, perfect wholeness, perfect adaptation, and in
the long run the Ascent of all living things.

This being so, the Law of the Struggle for Life is
elevated to a unique place in Nature as a first necessity of progress. It involves that every living thing
in nature shall live its best, that every resource shall
be called out to its utmost, that every individual
faculty shall be kept in the most perfect order and
work up to its fullest strength. So far from being a
drag on life, it is the one thing which not only makes
life go on at all, but which in the very act perfects it.
The result may sometimes involve the dethroning of a
species, or its entire extinction : it may lead in the
case of others to degeneration ; but in the end it must

result in the gradual perfecting of organisms upon the whole, and the steady advance of the final type. In fixing the eye on the murderous side of this Struggle, it is therefore well to remember to what it leads. There could be no higher end in the universe than to make a perfect world, and no more perfect law than that which at the same moment eliminates the unfit and establishes the fit. Too frequently the moralist's attention is diverted to the negative side, to what seems the quite immoral spectacle of the massacre of the innocent, the rout and murder of the unfit. But in earlier Nature there is no such word as innocent; and no ethical meaning at that stage can attach itself to the term " unfit." Fitness in the stormy days of the world's animal youth was necessarily fighting-fitness; no higher end was present anywhere than simply to gain for life a footing in the world, and perfect it up to the highest physical form. The creature which did that fulfilled its destiny, and no higher destiny was possible or conceivable. The Survival of the Fittest, of course, does not mean the survival of the strongest. It means the Survival of the Adapted —the survival of the most fitted to the circumstances which surround it. A fish survives in water when a leaking ironclad goes to the bottom, not because it is stronger but because it is better adapted to the element in which it lives. A Texas bull is stronger than a mosquito, but in an autumn drought the bull dies, the mosquito lives. Fitness to survive is simply fittedness, and has nothing to do with strength or courage, or intelligence or cunning as such, but only with adjustments as fit or unfit to the world around. A prize-fighter is stronger than a cripple; but in the

14

environment of modern life the cripple is cared for by
the people, is judged fit to live by a moral world,
while the pugilist, handicapped by his very health,
has to conduct his own struggle for existence. Physical fitness here is actually a disqualification ; what
was once unfitness is now fitness to survive. As we
rise in the scale, the physical fitness of the early world
changes to fitness of a different quality, and this law
becomes the guardian of a moral order. In one era
the race is to the swift, in another the meek inherit
the earth. In a material world social survival depends on wealth, health, power ; in a moral world the
fittest are the weak, the pitiable, the poor. Thus
there comes a time when this very law, in securing
survival for those who would otherwise sink and fall,
is the minister of moral ends.

When we pass from the animal and the savage
states to watch the working of the Struggle for Life
in later times, the impression deepens that after all,
the " gladiatorial theory " of existence has much to
say for itself. To trace its progress further is denied
us for the present, but observe before we close what
it connotes in modern life. Its lineal descendants are
two in number, and they have but to be named to
show the enormous place this factor has been given
to play in the world's destiny. The first is War, the
second is Industry. These in all their forms and
ramifications are simply the primitive Struggle continued on the social and political plane. War is not a
casual thing like a thunderstorm, nor a specific thing
like a battle. It is that ancient Struggle for Life carried over from the animal kingdom, which, in the later
as in the earlier world, has been so perfect an instru-

ment of evolution. Along with Industry, and for a time before it, War was the foster-mother of civilization. The patron of the heroic virtues, the purifier of societies, the solidifier of states, the military form of this Struggle—despite the awful balance on the other side—stands out on every page of history as the maker and educator of the human race. Industry is but the same Struggle in another disguise. The industrial conflict of to-day is the old attempt of primitive Man to get the most out of Nature—to grow foods, to find clothes, to raise fuel, to gain wealth. Owing to the ever-increasing number of the Strugglers the supplies fall short of the demands, with the result of perpetuating on the industrial plane, and often in hard and degrading forms, the primitive Struggle for Life. When society wonders at its labor troubles it forgets that Industry is a stage but one or two removes from the purely animal Struggle. And when morality impugns the Struggle for Life, it forgets that nearly the whole later fabric of civilization is its creation.

But one has only to look at these further phases of the Struggle to observe the most important fact of all—the change that passes over the principle as time goes on. Examine it on the higher levels as carefully as we have examined it on the lower, and though the crueler elements persist with fatal and appalling vigor, there are whole regions, and daily enlarging regions, where every animal feature is discredited, discouraged, or driven away. Already, with the social tragedy still at its height around us, the amelioration in many directions makes constant progress; and partly through the rise of opposing forces, and partly through the very civilization which it has helped to create, the maligner power must disappear. The Struggle for Life, as life's dynamic, can never wholly cease. In the keenness of its energies, the splendor of its stimulus, its bracing effect on character, its wholesome tensions throughout the whole range of action, it must remain with us to the end. But in the virulence of its animal qualities it must surely pass away. There are those who, without reflecting on this qualitative change, would govern Society by the merely animal Struggle ; those who claim for this the sanction of Nature, and lay down the principle of selfishness as the eternally working law. The eternal law, as we shall presently see, is unselfishness. But even the selfishness of early Nature loses its sting with time ; the self that is in it becomes a higher self ; and the world in which it acts is so much a better world that if self gave full rein to the animal it would be instantly extinguished.

The amelioration of the Struggle for Life is the most certain prophecy of Science. If this universe is a moral universe, it was a necessity that sooner or later this conflict should abate, that in the course of Evolution this particular change should come, that there should be put into the very machinery of Nature that which should bring it about. And what do we find ? We find the Animal side of the Struggle for Life attacked in such directions, and with such weapons that its defeat is sure. These weapons are in the armory of Nature ; they have been there from the beginning; and they are now engaged upon the enemy so hotly and so openly that we can discover what some of them are. The first is one which has begun to mine the Struggle for Life at its roots. Essentially, as we have seen, the Struggle for Life is the attempt to solve the fundamental problem of all life —Nutrition. If that could be solved apart from the Struggle for Life, its occupation would be gone. Now, it is more than probable that that problem will be otherwise solved. It will be solved by science. At the present moment Chemistry is devoting itself to the experiment of *manufacturing nutrition*, and with an enthusiasm which only immediate hope begets. It is not the visionaries who have dared to prophesy here. In a hundred laboratories the problem is being practically worked out, and, as one of the highest authorities assures us, " The time is not far distant when the artificial preparation of articles of food will be accomplished."[1] Already, through the labors of other sciences, the Struggle for Food has been made infinitely easier than it was ; but when the immediate quest succeeds, and the food of Man is made direct from the elements, the Struggle in all its coarser forms will practically be abolished. Civilization cannot ease the whole burden at once; the Struggle for Life will go on, but it will be the Struggle with its fangs drawn.

But there is a higher hope than Science. Attacked from below by Man's intellect, the final blow will be struck from a deeper source. It is impossible to conceive that the Ascent of Man should always depend upon his appetites, that in God's world there should be nothing better to attract him than food and raiment, that he should take no single step towards a higher life except when driven to it. As there comes

[1] Prof. Remsen, *M'Clure's Magazine*, Jan., 1804.

a time in a child's life when coercion gives place to free and conscious choice, the day comes to the world when the aspirations of the spirit begin to compete with, to neutralize, and to supplant the compulsions of the body. Against that day in the heart of humanity, Nature had made full provision. For there, prepared by a profounder chemistry than that which was to relieve the strain on the physical side, had gathered through the ages a force in whose presence the energies of the Animal Struggle are as naught. Beside the *Struggle for the Life of Others* the Struggle for Life is but a passing phase. As old, as deeply sunk in Nature, this further force was destined from the first to replace the Struggle for Life, and to build a nobler superstructure on the foundations which it laid. To establish these foundations was all that the Animal Struggle was ever designed to do. It has laid them well; yet it is only when the Struggle for Life stands projected against the larger influence with which all through history—and in an infinitely profound sense through moral history—it has been allied, that at once its worth and its ignominy are seen.

CHAPTER VII.

THE STRUGGLE FOR THE LIFE OF OTHERS.

WE now open a wholly new, and by far the most important, chapter in the Evolution of Man. Up to this time we have found for him a Body, and the rudiments of Mind. But Man is not a Body, nor a Mind. The temple still awaits its final tenant—the higher human Soul.

With a Body alone, Man is an animal: the highest animal, yet a pure animal; struggling for its own narrow life, living for its small and sordid ends. Add a Mind to that and the advance is infinite. The Struggle for Life assumes the august form of a struggle for light: he who was once a savage, pursuing the arts of the chase, realizes Aristotle's ideal man, "a hunter after Truth." Yet this is not the end. Experience tells us that Man's true life is neither lived in the material tracts of the body, nor in the higher altitudes of the intellect, but in the warm world of the affections. Till he is equipped with these Man is not human. He reaches his full height only when Love becomes to him the breath of life, the energy of will, the summit of desire. There at last lies all happiness, and goodness, and truth and divinity:

" For the loving worm within its clod
  Were diviner than a loveless God."

That Love did not come down to us through the
Struggle for Life, the only great factor in Evolution
which up to this time has been dwelt upon, is self-evi-
dent. It has a lineage all its own. Yet inexplicable
though the circumstance be, the history of this force,
the most stupendous the world has ever known, has
scarcely even begun to be investigated. Every other
principle in Nature has had a thousand prophets; but
this supreme dynamic has run its course through the
ages unobserved; its rise, so far as science is concerned, is unknown; its story has never been told.
But if any phenomenon or principle in Nature is capable of treatment under the category of Evolution, this
is. Love is not a late arrival, an after-thought, with
Creation. It is not a novelty of a romantic civilization. It is not a pious word of religion. Its roots began to grow with the first cell of life which budded on
this earth. How great it is, the history of humanity
bears witness : but how old it is and how solid, how
bound up with the very constitution of the world,
how from the first of time an eternal part of it, we are
only now beginning to perceive. For the Evolution of
Love is a piece of pure Science. Love did not descend
out of the clouds like rain or snow. It was distilled
on earth. And few of the romances which in after
years were to cluster round this immortal word are
more wonderful than the story of its birth and
growth. Partly a product of crushed lives and exter-
minated species, and partly of the choicest blossoms
and sweetest essences that ever came from the tree of
life, it reached its spiritual perfection after a history

the most strange and checkered that the pages of Nature have to record. What Love was at first, how crude and sour and embryonic a thing, it is impossible to conceive. But from age to age, with immeasurable faith and patience, by cultivations continuously repeated, by transplantings endlessly varied, the unrecognizable germ of this new fruit was husbanded to its maturity, and became the tree on which humanity, society, and civilization were ultimately borne.

As the story of Evolution is usually told, Love—the evolved form, as we shall see, of the Struggle for the Life of Others—has not even a place. Almost the whole emphasis of science has fallen upon the opposite—the animal Struggle for Life. Hunger was early seen by the naturalists to be the first and most imperious appetite of all living things, and the course of Nature came to be erroneously interpreted in terms of a never-ending strife. Since there are vastly more creatures born than can ever survive, since for every morsel of food provided a hundred claimants appear, life to an animal was described to us as one long tragedy; and Poetry, borrowing the imperfect creed, pictured Nature only as a blood-red fang. Before we can go on to trace the higher progress of Love itself, it is necessary to correct this misconception. And no words can be thrown away if they serve, in whatever imperfect measure, to restore to honor what is in reality the supreme factor in the Evolution of the world. To interpret the whole course of Nature by the Struggle for Life is as absurd as if one were to define the character of St. Francis by the tempers of his childhood. Worlds grow up as well as infants; their tempers change, the better nature opens out, new objects of desire appear, higher activities are added to the lower. The first chapter or two of the story of Evolution may be headed the Struggle for Life; but take the book as a whole and it is not a tale of battle. It is a Love-story.

The circumstances, as has been already pointed out in the Introduction, under which the world at large received its main impression of Evolution, obscured these later and happier features. The modern revival of the Evolution theory occurred almost solely in connection with investigations in the lower planes of Nature, and was due to the stimulus of the pure naturalists, notably of Mr. Darwin. But what Mr. Darwin primarily undertook to explain was simply the Origin of Species. His work was a study in infancies, in rudiments; he emphasized the earliest forces and the humblest phases of the world's development. The Struggle for Life was there the most conspicuous fact—at least, on the surface; it formed the key-note of his teaching; and the tragic side of Nature fixed itself in the popular mind. The mistake the world made was twofold: it mistook Darwinism for Evolution—a specific theory of Evolution applicable to a single department, for a universal scheme; and it misunderstood Mr. Darwin himself. That the foundations of Darwinism—or what was taken for Darwinism—were the foundations of all Nature was assumed. Dazzled with the apparent solidity of this foundation, men made haste to run up a structure which included the whole vast range of life—vegetal, animal, social—based on a law which explained but half the facts, and was only relevant, in the crude form in which it was universally stated, for the childhood of the world. It was impossible for such an edifice to stand. Natural history cannot in any case cover the whole facts of human history, and, so interpreted, can only fatally distort them. The mistake had been largely qualified had Mr. Darwin's followers even accepted his foundation in its first integrity; but, perhaps because the author of the theory himself but dimly apprehended the complement of his thesis, few seem to have perceived that anything was amiss. Mr. Darwin's sagacity led him distinctly to foresee that narrow interpretations of his great phrase "Struggle for Existence" were certain to be made; and in the opening chapters of the *Origin of Species,* he warns us that the term must be applied in its "large and metaphorical sense, including dependence of one being on another, and including (which is more important) not only the life of the individual, but success in leaving progeny." [1] In spite of this warning, his overmastering emphasis on the individual Struggle for Existence seems to have obscured, if not to his own mind, certainly to almost all his followers, the truth that any other great factor in Evolution existed.

The truth is there are *two* Struggles for Life in every living thing—the Struggle for Life, and the Struggle for the Life of Others. The web of life is woven upon a double set of threads, the second thread distinct in color from the first, and giving a totally different pattern to the finished fabric. As the whole aspect of the after-world depends on this distinction of strands in the warp, it is necessary to grasp the distinction with the utmost clearness. Already, in the introductory chapter, the nature of the distinction

[1] *Origin of Species*, 6th edition, p. 50.

has been briefly explained. But it is necessary to be explicit here, even to redundancy. We have arrived at a point from which the Ascent of Man takes a fresh departure, a point from which the course of Evolution begins to wear an entirely altered aspect. No such consummation ever before occurred in the progress of the world as the rise to potency in human life of the Struggle for the Life of Others. The Struggle for the Life of Others is the physiological name for the greatest word of ethics—Other-ism, Altruism, Love. From Self-ism to Other-ism is the supreme transition of history. It is therefore impossible to lodge in the mind with too much solidity the simple biological fact on which the Altruistic Struggle rests. Were this a late phase of Evolution, or a factor applicable to single genera, it would still be of supreme importance ; but it is radical, universal, involved in the very nature of life itself. As matter is to be interpreted by Science in terms of its properties, life is to be interpreted in terms of its functions. And when we dissect down to that form of matter with which all life is associated, we find it already discharging in the humblest organisms visible by the microscope the function on which the stupendous superstructure of Altruism indirectly comes to rest. Take the tiniest protoplasmic cell, immerse it in a suitable medium, and presently it will perform two great acts—the two which sum up life, which constitute the eternal distinction between the living and the dead—Nutrition and Reproduction. At one moment, in pursuance of the Struggle for Life, it will call in matter from without, and assimilate it to itself ; at another moment, in pursuance of the Struggle for the Life of Others, it will set a portion of

that matter apart, add to it, and finally give it away
to form another life. Even at its dawn life is receiver
and giver ; even in protoplasm is Self-ism and Other-
ism. These two tendencies are not fortuitous. They
have been lived into existence. They are not grafts
on the tree of life, they are its nature, its essential
life. They are not painted on the canvas, but woven
through it.

The two main activities, then, of all living things are
Nutrition *and Reproduction.* The discharge of these
functions in plants, and largely in animals, sums up
the work of life. The object of Nutrition is to secure
the life of the individual ; the object of Reproduction
is to secure the life of the Species. These two objects
are thus wholly different. The first has a purely personal end ; its attention is turned inwards ; it exists
only for the present. The second in a greater or less
degree is impersonal ; its attention is turned out-
wards ; it lives for the future. One of these objects,
in other words, is Self-regarding ; the other is Other-
regarding. Both, of course, at the outset are wholly
selfish ; both are parts of the Struggle for Life. Yet
see already in this non-ethical region a parting of the
ways. Selfishness and unselfishness are two supreme
words in the moral life. The first, even in physical
Nature, is accompanied by the second. In the very
fact that one of the two mainsprings of life is Other-
regarding there lies a prophecy, a suggestion, of the
day of Altruism. In organizing the physiological
mechanism of Reproduction in plants and animals
Nature was already laying wires on which, one far-off
day, the currents of all higher things might travel.

In itself, this second struggle, this effort to main-

tain the life of the species, is not less real than the
first; the provisions for effecting it are not less wonderful; the whole is not less a part of the system of
things. And, taken prophetically, the function of
Reproduction is as much greater than the function of
Nutrition as the Man is greater than the Animal, as
the Soul is higher than the Body, as Co-operation is
stronger than Competition, as Love is stronger than
Hate. If it were ever to be charged against Nature
that she was wholly selfish, here is the refutation
at the very start. One of the two fundamental activities of all life, whether plant or animal, is Other-regarding. It is not said that the function of Reproduction, say in a fern or in an oak, is an unselfish act,
yet in a sense, even though begotten of self, it is an
other-regarding act. In the physical world, to speak
of the Struggle for Food as selfish, or to call the Struggle for Species unselfish, are alike incongruous. But
if the morality of Nature is impugned on the ground
of the universal Struggle for Life, it is at least as relevant to refute the charge by putting moral content
into the universal Struggle for Species. No true
moral content can be put into either, yet the one
marks the beginning of Egoism, the other of Altruism.
Almost the whole self-seeking side of things has come
down the line of the individual Struggle for Life; almost the whole unselfish side of things is rooted in the
Struggle to preserve the life of others.

That an Other-regarding principle should sooner or
later appear on the world's stage was a necessity if
the world was ever to become a moral world. And as
everything in the moral world has what may be called
a physical basis to begin with, it is not surprising to

find in the mere physiological process of Reproduction a physical forecast of the higher relations, or, more accurately, to find the higher relations manifesting themselves at first through physical relations. The Struggle for the Life of Others formed an indispensable stepping-stone to the development of the Other-regarding virtues. Nature always works with long roots. To conduct Other-ism upward into the higher sphere without miscarriage, and to establish it there forever, Nature had to embed it in the most ancient past, so organizing and endowing protoplasm that life could not go on without it, and compelling its continuous activity by the sternest physiological necessity.

To say that there is a certain protest of the mind against associating the highest ethical ends with forces in their first stage almost physical, is to confess a truth which all must feel. Even Hæckel, in contrasting the tiny rootlet of sex-attraction between two microscopic cells with the mighty after-efflorescence ot love in the history of mankind, is staggered at the audacity of the thought, and pauses in the heart of a profound scientific investigation to reflect upon it. After a panegyric in which he says, " We glorify love as the source of the most splendid creations of art ; of the noblest productions of poetry, of plastic art, and of music ; we reverence in it the most powerful factor in human civilization, the basis of family life, and, consequently, of the development of the state ; " . . . he adds, " So wonderful is love, and so immeasurably important is its influence on mental life, that in this point, more than in any other, ' supernatural ' causation seems to mock every natural explanation." It is the mystery of Nature, that between the loftiest spiritual heights, and the lowliest physical depths, there should seem to run a pathway which the intellect of Man may climb. Haeckel has spoken, and rightly, from the stand-point of humanity; yet he continues, and with equal right, from the stand-point of the naturalist. " Notwithstanding all this, the comparative history of evolution leads us back very clearly and indubitably to the oldest and simplest source of love, to the elective affinity of two differing cells." [1]

SELF-SACRIFICE IN NATURE.

It is not, however, in Haeckel's " elective affinity of differing cells " that we must seek the physical basis of Altruism. That may be the physical basis of a passion which is frequently miscalled Love; but Love itself, in its true sense as Self-sacrifice, Love with all its beautiful elements of sympathy, tenderness, pity, and compassion, has come down a wholly different line. It is well to be clear about this at once, for the function of Reproduction suggests to the biological mind a view of this factor which would limit its action to a sphere which in reality forms but the merest segment of the whole. The Struggle for the Life of Others has certainly connected with it sex-relations, as we shall see; but we can only use it scientifically in its broad physiological sense, as literally a Struggling for Others, a giving up self for Others. And these others are not Other-sexes. They have nothing to do with sex. They are the fruits of Reproduction—the egg, the seed, the nestling, the little child. So far from its chief manifestation being

[1] Haeckel, *Evolution of Man*, Vol. II., p. 394.

within the sphere of sex it is in the care and nurt‧ure of the young, in the provision everywhere throughout Nature for the seed and egg, in the endless and infinite self-sacrifices of Maternity. that Altruism finds its main expression.

That this is the true reading of the work of this second factor appears even in the opening act of Reproduction in the lowest plant or animal. Pledged by the first law of its being—the law of self-preservation—to sustain itself, the organism is at the same moment pledged by the second law to give up itself. Watch one of the humblest unicellular organisms at the time of Reproduction. The cell, when it grows to be a certain size, divides itself into two, and each part sets up an independent life. Why it does so is now known. The protoplasm inside the cell—the body as it were—needs continually to draw in fresh food. This is secured by a process of imbibition or osmosis through the surrounding wall. But as the cell grows large, there is not wall enough to pass in all the food the far interior needs, for while the bulk increases as the cube of the diameter; the surface increases only as the square. The bulk of the cell, in short, has outrun the absorbing surface; its hunger has outgrown its satisfactions; and unless the cell can devise some way of gaining more surface it must starve. Hence the splitting into two smaller cells. There is now more absorbing surface than the two had when combined. When the two smaller cells have grown as large as the original parent, income and expenditure will once more balance. As growth continues, the waste begins to exceed the power of repair and the life of the cell is again threatened.

15

The alternatives are obvious. It must divide, or die. If it divides, what has saved its life! Self-sacrifice. By giving up its life as an individual, it has brought forth two individuals and these will one day repeat the surrender. Here, with differences appropriate to their distinctive spheres, is the first great act of the moral life. All life, in the beginning, is self-contained, self-centred, imprisoned in a single cell. The first step to a more abundant life is to get rid of this limitation. And the first act of the prisoner is simply to break the walls of its cell. The plant does this by a mechanical or physiological process; the moral being by a conscious act which means at once the breaking-up of Self-ism and the recovery of a larger self in Altruism. Biologically, Reproduction begins as rupture. It is the release of the cell, full-fed, yet unsatiated, from itself. "Except a corn of wheat fall into the ground and die, it abideth alone; but if it die, it bringeth forth much fruit."

These facts are not colored to suit a purpose. There is no other language in which science itself can state them. "Reproduction begins as rupture. Large cells beginning to die, save their lives by sacrifice. Reproduction is literally a life-saving against the approach of death. Whether it be the almost random rupture of one of the more primitive forms such as *Schizogenes*, or the overflow and separation of multiple buds as in *Arcella*, or the dissolution of a few of the Infusorians, an organism, which is becoming exhausted, saves itself and multiplies in reproducing." [1] There is no Reproduction in plant, animal, or Man which does not involve self-sacrifice. All that is

[1] *The Evolution of Sex*, page 232.

moral, and social, and other-regarding has come along
the line of this function. Sacrifice, moreover, as these
physiological facts disclose, is not an accident, nor
an accompaniment of Reproduction, but an inevitable
part of it. It is the universal law and the universal
condition of life. The act of fertilization is the
anabolic restoration, renewal, and rejuvenescence of
a katabolic cell : it is a resurrection of the dead
brought about by a sacrifice of the living, a dying of
part of life in order to further life.

Pass from the unicellular plant to one of the higher
phanerogams, and the self-sacrificing function is seen
at work with still greater definiteness, for there we
have a clearer contrast with the other function. To
the physiologist a tree is not simply a tree, but a complicated piece of apparatus for discharging, in the first
place, the function of Nutrition. Root, trunk, branch,
twig, leaf, are so many organs—mouths, lungs, cir-
culatory-system, alimentary canal—for carrying on to
the utmost perfection the Struggle for Life. But this
is not all. There is another piece of apparatus within
this apparatus of a wholly different order. It has
nothing to do with Nutrition. It has nothing to do
with the Struggle for Life. It is the flower. The
more its parts are studied, in spite of all homol-
ogies, it becomes more clear that this is a construction of a unique and wonderful character. So important has this extra apparatus seemed to science,
that it has named the great division of the vegetable
kingdom to which this and all higher plants belong,
the Phanerogams—the flowering plants ; and it
recognizes the complexity and physiological value of
this reproductive specialty by giving them the place

of honor at the top of the vegetable creation. Watch this flower at work for a little, and behold a miracle. Instead of struggling for life it lays down its life. After clothing itself with a beauty which is itself the minister of unselfishness, it droops, it wastes, it lays down its life. The tree still lives; the other leaves are fresh and green; but this life within a life is dead. And why? Because within this death is life. Search among the withered petals, and there, in a cradle of cunning workmanship, are a hidden progeny of clustering seeds—the gift to the future which this dying mother has brought into the world at the cost of leaving it. The food she might have lived upon is given to her children, stored round each tiny embryo with lavish care, so that when they waken into the world the first helplessness of their hunger is met. All the arrangements in plant-life which concern the flower, the fruit, and the seed are the creations of the Struggle for the Life of Others.

No one, though science is supposed to rob all the poetry from Nature, reverences a flower like the biologist. He sees in its bloom the blush of the young mother; in its fading, the eternal sacrifice of Maternity. A yellow primrose is not to him a yellow primrose. It is an exquisite and complex structure added on to the primrose plant for the purpose of producing other primrose plants. At the base of the flower, packed in a delicate casket, lie a number of small white objects no larger than butterflies' eggs. These are the eggs of the primrose. Into this casket, by a secret opening, filmy tubes from the pollen grains— now enticed from their hiding-place on the stamens and clustered on the stigma—enter and pour their fertilizing fovilla through a microscopic gateway which opens in the wall of the egg and leads to its inmost heart. Mysterious changes then proceed. The embryo of a future primrose is born. Covered with many protective coats, it becomes a seed. The original casket swells, hardens, is transformed into a rounded capsule opening by valves or a deftly constructed hinge. One day this capsule, crowded with seeds, breaks open and completes the cycle of Reproduction by dispersing them over the ground. There, by and bye, they will burst their enveloping coats, protrude their tiny radicles, and repeat the cycle of their parents' sacrificial life.

With endless variations in detail, these are the closing acts in the Struggle for the Life of Others in the vegetable world. We have illustrated the point from plants, because this is the lowest region where biological processes can be seen in action, and it is essential to establish beyond dispute the fundamental nature of the reproductive function. From this level onwards it might be possible to trace its influence, and growing influence, throughout the whole range of the animal kingdom until it culminates in its most consummate expression—a human mother. Some of the links in this unbroken ascent will be filled in at a later stage—for the Evolution of Maternity is so wonderful and so intricate as to deserve a treatment of its own—but meantime we must pass on to notice a few of the other gifts which Reproduction has bestowed upon the world. In a rigid sense, it is impossible to separate the gains to humanity from the Reproductive function as distinguished from those of the Nutritive. They are co-operators, not competitors, and their apparently rival paths continuously intertwine. But mark a few of the things that have mainly grown up around this second function and decide whether or not it be a worthy ally of the Struggle for Life in the Evolution of Man.

To begin at the most remote circumference, consider what the world owes to-day to the Struggle for the Life of Others in the world of plants. This is the humblest sphere in which it can offer any gifts at all, yet these are already of such a magnitude that without them the higher world would not only be inexpressibly the poorer, but could not continue to exist. As we have just seen, all the arrangements in plant life which concern the flower are the creations of the Struggle for the Life of Others. For Reproduction alone the flower is created ; when the process is over it returns to the dust. This miracle of beauty is a miracle of Love. Its splendor of color, its variegations, its form, its symmetry, its perfume, its honey, its very texture, are all notes of Love—Love-calls or Love-lures or Love-provisions for the insect world, whose aid is needed to carry the pollen from anther to stigma, and perfect the development of its young. Yet this is but a thing thrown in, in giving something else. The Flower, botanically, is the herald of the Fruit. The Fruit, botanically, is the cradle of the Seed. Consider how great these further achievements are, how large a place in the world's history is filled by these two humble things—the Fruits and Seeds of plants. Without them the Struggle for Life itself would almost cease. The animal Struggle for Life is **a struggle** for what ? For Fruits and Seeds. All animals in the long run depend for food upon **Fruits**

and Seeds, or upon lesser creatures which have utilized Fruits and Seeds. Three-fourths of the population of the world at the present moment subsist
upon rice. What is rice? It is a seed; a product of
Reproduction. Of the other fourth, three-fourths
live upon grains—barley, wheat, oats, millet. What
are these grains? Seeds—stores of starch or albumen
which, in the perfect forethought of Reproduction,
plants bequeath to their offspring. The foods of the
world, especially the children's foods, are the foods
of the children of plants, the foods which unselfish
activities store round the cradles of the helpless, so
that when the sun wakens them to their new world
they may not want. Every plant in the world lives
for Others. It sets aside something, something costly,
cared for, the highest expression of its nature. The
Seed is the tithe of Love, the tithe which Nature
renders to Man. When Man lives upon Seeds he
lives upon Love. Literally, scientifically, Love is Life.
If the Struggle for Life has made Man, braced and
disciplined him, it is the Struggle for Love that sustains him.

Pass from the foods of Man to drinks, and the gifts
of Reproduction once more all but exhaust the list.
This may be mere coincidence, but a coincidence
which involves both food and drink is at least worth
noting. The first and universal food of the world is
milk, a product of Reproduction. All distilled spirits
are products of Reproduction. All malted liquors are
made from the embryos of plants. All wines are
juices of the grape. Even on the plane of the animal
appetites, in mere relation to Man's hunger and his
thirst, the factor of Reproduction is thus seen to be

fundamental. To interpret the course of Evolution without this would be to leave the richest side even of material Nature without an explanation. Retrace the ground even thus hastily travelled over, and see how full Creation is of meaning, of anticipation, of good for Man, how far back begins the undertone of Love. Remember that nearly all the beauty of the world is Love-beauty—the corolla of the flower and the plume of the grass, the lamp of the fire-fly, the plumage of the bird, the horn of the stag, the face of a woman; that nearly all the music of the natural world is Love-music—the song of the nightingale, the call of the mammal, the chorus of the insect, the serenade of the lover; that nearly all the foods of the world are Love-foods—the date and the raisin, the banana and the bread-fruit, the locust and the honey, the eggs, the grains, the seeds, the cereals, and the legumes; that all the drinks of the world are Love-drinks—the juices of the sprouting grain and the withered hop, the milk from the udder of the cow, the wine from the Love-cup of the vine. Remember that the Family, the crown of all higher life, is the creation of Love; that Co-operation, which means power, which means wealth, which means leisure, which therefore means art and culture, recreation and education, is the gift of Love. Remember not only these things, but the diffusions of feeling which accompany them, the elevations, the ideals, the happiness, the goodness, and the faith in more goodness, and ask if it is not a world of Love in which we live.

CO-OPERATION IN NATURE.

Though Co-operation is not exclusively the gift of

Reproduction, it is so closely related to it that we may next observe a few of the fruits of this most definitely altruistic principle. For here is a principle, not merely a series of interesting phenomena, profoundly rooted in Nature and having for its immediate purpose the establishment of Other-ism. In innumerable cases, doubtless, Co-operation has been induced rather by the action of the Struggle for Life— a striking circumstance in itself, as showing how the very selfish side of life has had to pay its debt to the larger law—but in multitudes more it is directly allied with the Struggle for the Life of Others.

For illustrations of the principle in general we may begin with the very dawn of life. Every life at first was a single cell. Co-operation was unknown. Each cell was self-contained and self-sufficient, and as new cells budded from the parent they moved away and set up life for themselves. This self-sufficiency leads to nothing in Evolution. Unicellular organisms may be multiplied to infinity, but the vegetable kingdom can never rise in height, or symmetry, or productiveness without some radical change. But soon we find the co-operative principle beginning its mysterious integrating work. Two, three, four, eight, ten cells club together and form a small mat, or cylinder, or ribbon—the humblest forms of corporate plant-life—in which each individual cell divides the responsibilities and the gains of living with the rest. The colony succeeds; grows larger; its co-operations become more close and varied. Division of labor in new directions arises for the common good; leaves are organized for nutrition, and special cells for reproduction. All the organs increase in specialization;

and the time arrives when from cryptogams the plant-world bursts into flowers. A flower is organized for Co-operation. It is not an individual entity, but a commune, a most complex social system. Sepal, petal, stamen, anther, each has its separate rôle in the economy, each necessary to the other and to the life of the species as a whole. Mutual aid having reached this stage can never be arrested short of the extinction of plant-life itself.

Even after this stage, so triumphant is the success of the Co-operative Principle, that having exhausted the possibilities of further development within the vegetable kingdom, it overflowed these boundaries and carried the activities of flowers into regions which the plant-world never invaded before. With a novelty and audacity unique in organic Nature, the higher flowering plants, stimulated by Co-operation, opened communication with two apparently forever unrelated worlds, and established alliances which secured from the subjects of these distant states, a perpetual and vital service. The history of these relations forms the most entrancing chapter in botanical science. But so powerfully has this illustration of the principle appealed already to the popular imagination, that it becomes a mere form to restate it. What interests us anew in these novel enterprises, nevertheless, is that they are directly connected with the Reproductive Struggle. For it is not for food that the plant-world voyages into foreign spheres, but to perfect the supremer labor of its life.

The vegetable world is a world of still life. No higher plant has the power to move to help its neighbor, or even to help itself, at the most critical

moment of its life. And it is through this very helplessness that these new Co-operations are called forth.
The fertilizing pollen grows on one part of the flower,
the stigma which is to receive it grows on another, or it
may be on a different plant. But as these parts can
not move towards one another, the flower calls in the
aid of moving things. Unconscious of their vicarious
service, the butterfly and the bee, as they flit from
flower to flower, or the wind as it blows across the
fields, carry the fertilizing dust to the waiting stigma,
and complete that act without which in a generation
the species would become extinct. No flower in the
world, at least no entomophilous flower, can continuously develop healthy offspring without the Co-operations of an insect; and multitudes of flowers without
such aid could never seed at all. It is to these Co-
operations that we owe all that is beautiful and
fragrant in the flower-world. To attract the insect
and recompense it for its trouble, a banquet of honey
is spread in the heart of the flower; and to enable the
visitor to find the nectar, the leaves of the flower are
made showy or conspicuous beyond all other leaves.
To meet the case of insects which love the dusk, many
flowers are colored white; for those which move
about at night and cannot see at all, the night-flowers
load the darkness with their sweet perfume. The
loveliness, the variegations of shade and tint, the
ornamentations, the scents, the shapes, the sizes of
flowers, are all the gifts of Co-operation. The flower
in every detail, in fact, is a monument to the Co-operative Principle.

Scarcely less singular are the Co-operations among
flowers themselves the better to attract the attention

of the insect world. Many flowers are so small and inconspicuous that insects might not condescend to notice them. But Altruism is always inventive. Instead of dispersing their tiny florets over the plant, these club together at single points, so that by the multitude of numbers an imposing show is made. Each of the associating flowers in these cases preserves its individuality, and—as we see in the Elder or the Hemlock—continues to grow on its own flower stalk. But in still more ingenious species the partners to a floral advertisement sacrifice their separate stems and cluster close together on a common head. The Thistle, for example, is not one flower, but a colony of flowers, each complete in all its parts, but all gaining the advantage of conspicuousness by densely packing themselves together. In the Sun-flowers and many others the sacrifice is carried still further. Of the multitude of florets clustered together to form the mass of color, a few cease the development of the reproductive organs altogether, and allow their whole strength to go towards adding visibility to the mass. The florets in the centre of the group, packed close together, are unable to do anything in this direction; but those on the margin expand the perianth into a blazing circle of flame, and leave the deep work of Reproduction to those within. What are the advantages gained by all this mutual aid? That it makes them the fittest to survive. These Co-operative Plants are among the most numerous, most vigorous, and most widely diffused in Nature. Self-sacrifice and Co-operation are thus recognized as sound in principle. The blessing of Nature falls upon them. The words themselves, in any more than a merely physical sense, are hopelessly out of court in any scientific interpretation of things. But the point to mark is that on the mechanical equivalent of what afterwards come to have ethical relations Natural Selection places a premium. Non-co-operative or feebly co-operative organisms go to the wall. Those which give mutual aid survive and people the world with their kind. Without pausing to note the intricate Co-operations of flowers which reward the eye of the specialist—the subtle alliance with Space in Diœcious flowers; with Time in Dichogamous species, and with Size in the Dimorphic and Trimorphic forms —consider for a moment the extension of the principle to the Seed and Fruit. Helpless, singlehanded, as is a higher plant, with regard to the efficient fertilizing of its flowers, an almost more difficult problem awaits it when it comes to the dispersal of its seeds. If each seed fell where it grew, the spread of the species would shortly be at an end. But Nature, working on the principle of Co-operation, is once more redundant in its provisions. By a series of new alliances the offspring are given a start on distant and unoccupied ground; and so perfect are the arrangements in this department of the Struggle for the Life of Others that single plants, immovably rooted in the soil, are yet able to distribute their children over the world. By a hundred devices the fruits and seeds when ripe are entrusted to outside hands—provided with wing or parachute and launched upon the wind, attached by cunning contrivances to bird and beast, or dropped into stream and wave and ocean-current, and so transported over the earth.

If we turn to the Animal Kingdom, the Principle of

Co-operation everywhere once more confronts us. It is singular that, with few exceptions, science should still know so little of the daily life of even the common animals. A few favorite mammals, some birds, three or four of the more picturesque and clever of the insects—these almost exhaust the list of those whose ways are thoroughly known. But, looking broadly at Nature, one general fact is striking—the more social animals are in overwhelming preponderance over the unsocial. Mr. Darwin's dictum, that "those communities which included the greatest number of the most sympathetic members would flourish best," is wholly proved. Run over the names of the commoner or more dominant mammals, and it will be found that they are those which have at least a measure of sociability. The cat-tribe excepted, nearly all live together in herds or troops—the elephant, for instance, the buffalo, deer, antelope, wild-goat, sheep, wolf, jackal, reindeer, hippopotamus, zebra, hyena, and seal. These are *mammals*, observe—an association of sociability in its highest developments with reproductive specialization. Cases undoubtedly exist where the sociability may not be referable primarily to this function ; but in most the chief Co-operations are centred in Love. So advantageous are all forms of mutual service that the question may be fairly asked whether after all Co-operation and Sympathy—at first instinctive, afterwards reasoned—are not the greatest facts even in organic Nature ? To quote the words of Prince Kropotkin : " As soon as we study animals— not in laboratories and museums only, but in the forest and the prairie, in the steppes and the mountains —we at once perceive that though there is an immense amount of warfare and extermination going on amidst various species, and especially amidst various classes of animals, there is, at the same time, as much, or perhaps more, of mutual support, mutual aid, and mutual defence, amidst animals belonging to the same species or, at least, to the same society. Sociability is as much a law of Nature as mutual struggle. . . . If we resort to an indirect test and ask Nature ' Who are the fittest: those who are continually at war with each other, or those who support one another?' we at once see that those animals which acquire habits of mutual aid are undoubtedly the fittest. They have more chances to survive, and they attain, in their respective classes, the highest development of intelligence and bodily organization. If the numberless facts which can be brought forward to support this view are taken into account, we may safely say that mutual aid is as much a law of animal life as mutual struggle ; but that, as a factor of evolution, it most probably has a far greater importance, inasmuch as it favors the development of such habits and character as insure the maintenance and further development of the species, together with the greatest amount of welfare and enjoyment of life for the individual, with the least waste of energy." [1]

In the large economy of Nature, almost more than within these specific regions, the inter-dependence of part with part is unalterably established. The system of things, from top to bottom, is an uninterrupted series of reciprocities. Kingdom corresponds with kingdom, organic with inorganic. Thus, to carry on the larger agriculture of Nature, myriads of living

[1] *Nineteenth Century*, 1890, p. 340.

creatures have to be retained in the earth itself—*in the earth*—and to prepare and renew the soils in which the otherwise exhausted ground may keep up her continuous gifts of vegetation. Ages before Man appeared with his tools of husbandry, these agriculturists of Nature—in humid countries the Worm, in subtropical regions the White Ant—ploughed and harrowed the earth, so that without the Co-operations of these most lowly forms of life, the higher beauty and fruitfulness of the world had been impossible. The very existence of animal life, to take another case of broad economy, is possible only through the mediation of the plant. No animal has the power to satisfy one single impulse of hunger without the Co-operation of the vegetable world. It is one of the mysteries of organic chemistry that the Chlorophyll contained in the green parts of plants, alone among substances, has the power to break up the mineral kingdom and utilize the products as food. Though detected recently in the tissues of two of the very lowest animals, Chlorophyll is the peculiar possession of the vegetable kingdom, and forms the solitary point of contact between Man and all higher animals and their supply of food. Every grain of matter therefore eaten by Man, every movement of the body, every stroke of work done by muscle or brain, depends upon the contribution of a plant, or of an animal which has eaten a plant. Remove the vegetable kingdom, or interrupt the flow of its unconscious benefactions, and the whole higher life of the world ends. Everything, indeed, came into being because of something else, and continues to be because of its relations to something else. The matter of the earth is built up of co-operating

atoms; it owes its existence, its motion, and its stability to co-operating stars. Plants and animals are
made of co-operating cells, nations of co-operating
men. Nature makes no move, Society achieves no
end, the Cosmos advances not one step, that is not dependent on Co-operation; and while the discords of
the world disappear with growing knowledge, Science
only reveals with increasing clearness the universality
of its reciprocities.

But to return to the more direct effects of Reproduction. After creating Others there lay before
Evolution a not less necessary task—the task of
uniting them together. To create units in indefinite quantities and scatter them over the world
is not even to take one single step in progress.
Before any higher evolution can take place these
units must by some means be brought into relation
so as not only to act together, but to react upon
each other. According to well-known biological
laws, it is only in combinations, whether of atoms,
cells, animals, or human beings, that individual
units can make any progress, and to create such
combinations is in every case the first condition
of development. Hence the first commandment
of Evolution everywhere is "Thou shalt mass,
segregate, combine, grow large." Organic Evolution, as Mr. Herbert Spencer tells us, "is primarily the formation of an aggregate." No doubt the
necessities of the Struggle for Life tended in many
ways to fulfil this condition, and the organization
of primitive societies, both animal and human, are
largely its creation. Under its influence these were
called together for mutual protection and mutual help;

16

and Co-operations induced in this way have played an important part in Evolution. But the Co-operations brought about by Reproduction are at once more radical, more universal, and more efficient. The Struggle for Life is in part a disruptive force. The Struggle for the Life of Others is wholly a social force. The social efforts of the first are secondary; those of the last are primary. And had it not been for the stronger and unbreakable bond which the Struggle for the Life of Others introduced into the world the organization of Societies had never even been begun. How subtly Reproduction effects its purpose an illustration will make plain. And we shall select it again from the lowest world of life, so that the fundamental nature of this factor may be once more vindicated on the way.

More than two thousand years ago Herodotus observed a remarkable custom in Egypt. At a certain season of the year, the Egyptians went into the desert, cut off branches from the wild palms, and, bringing them back to their gardens, waved them over the flowers of the date-palm. Why they performed this ceremony they did not know; but they knew that if they neglected it, the date crop would be poor or wholly lost. Herodotus offers the quaint explanation that along with these branches there came from the desert certain flies possessed of a " vivific virtue," which somehow lent an exuberant fertility to the dates. But the true rationale of the incantation is now explained. Palm-trees, like human beings, are male and female. The garden plants, the date-bearers, were females ; the desert plants were males ; and the waving of the branches over the females meant the transference of the fertilizing pollen dust from the one to the other.

Now consider, in this far-away province of the vegetable kingdom, the strangeness of this phenomenon. Here are two trees living wholly different lives, they are separated by miles of desert sand ; they are unconscious of one another's existence ; and yet they are so linked together that their separation into two is a mere illusion. Physiologically they are one tree ; they cannot dwell apart. It is nothing to the point that they are neither dowered with locomo tion nor the power of conscious choice. The point is that there is that in Nature which unites these seemingly disunited things, which effects combinations and co-operations where one would least believe them possible, which sustains by arrangements of the most elaborate kind inter-relations between tree and tree. By a device the most subtle of all that guard the higher Evolution of the world—the device of Sex —Nature accomplishes this task of throwing irre-sistible bonds around widely separate things, and establishing such sympathies between them that they must act together or forfeit the very life of their kind. Sex is a paradox; it is that which sepa-rates in order to unite. The same mysterious mesh which Nature threw over the two separate palms, she threw over the few and scattered units which were to form the nucleus of Mankind.

Picture the state of primitive Man ; his fear of other primitive Men ; his hatred of them ; his un-sociability ; his isolation ; and think how great a thing was done by Sex in merely starting the crystal-lization of humanity. At no period, indeed, was Man ever utterly alone. There is no such thing in nature as *a man*, or for the matter of that as *an* animal, except among the very humblest forms. Wherever there is a higher animal there is another animal; wherever there is a savage there is another savage— the other half of him, a female savage. This much, at least, Sex has done for the world—it has abolished the numeral *one*. Observe, it has not simply discouraged the existence of one; it has abolished the existence of one. The solitary animal must die, and can leave no successor. Unsociableness, therefore, is banished out of the world; it has become the very condition of continued existence that there should always be a family group, or at least pair. The determination of Nature to lay the foundation stone of corporate national life at this point, and to embed Sociability forever in the constitution of humanity, is only obvious when we reflect with what extraordinary thoroughness this Evolution of Sex was carried out. There is no instance in Nature of Division of Labor being brought to such extreme specialization. The two sexes were not only set apart to perform different halves of the same function, but each so entirely lost the power of performing the whole function that even with so great a thing at stake as the continuance of the species *one* could not discharge it. Association, combination, mutual help, fellowship, affection—things on which all material and moral progress would ultimately turn—were thus forced upon the world at the bayonet's point.

This hint, that the course of development is taking a social rather than an individual direction, is of immense significance. If that can be brought about by the Struggle for the Life of Others—and in the next chapters we shall see that it has been—there can be no dispute about the rank of the factor which consummates it. Along the line of the physiological function of Reproduction, in association with its induced activities and relations, not only has Altruism entered the world, but along with it the necessary field for its expansion and full expression. If Nature is to be read solely in the light of the Struggle for Life, these ethical anticipations—and as yet we are but at the beginning of them—for a social world and a moral life, must remain the stultification both of science and of teleology.

THE ETHICAL SIGNIFICANCE OF SEX.

Next among the gifts of Reproduction fall to be examined some further contributions yielded by the new and extraordinary device which a moment ago leaped into prominence—Sex. The direct, and especially the collateral, issues here are of such significance that it will be essential to study them in detail. Realize the novelty and originality of this most highly specialized creation, and it will be seen at once that something of exceptional moment must lie behind it. Here is a phenomenon which stands absolutely alone on the field of Nature. There is not only nothing at all like it in the world, but while everything else has homologues or analogues somewhere in the cosmos, this is without any parallel. Familiarity has so accustomed us to it that we accept the sex separation as a matter of course ; but no words can do justice to the wonder and novelty of this strange line of cleavage which cuts down to the very root of being in everything that lives.

No theme of equal importance has received less attention than this from evolutionary philosophy. The single problems which sex suggests have been investigated with a keenness and brilliance of treatment never before brought to bear in this mysterious region; and Mr. Darwin's theory of sexual selection, whether true or false, has called attention to a multitude of things in living Nature which seem to find a possible explanation here. But the broad, and simple fact that this division into maleness and femaleness should run between almost every two of every plant and every animal in existence, must have implications of a quite exceptional kind.

How deep, from the very dawn of life, this rent between the two sexes yawns is only now beginning to be seen. Examine one of the humblest water weeds—the Spirogyra. It consists of waving threads or necklaces of cells, each plant to the eye the exact duplicate of the other. Yet externally alike as they seem, the one has the physiological value of the male, the other of the female. Though a primitive method of Reproduction, the process in this case foreshadows the law of all higher vegetable life. From this point upwards, though there are many cases where reproduction is asexual, in nearly every family of plants a Reproduction by spores takes place, and where it does not take place its absence is abnormal, and to be accounted for by degeneration. When we reach the higher plants the differences of sex become as marked as among the higher animals. Male and female flowers grow upon separate trees, or live side by side

on the same branch, yet so unlike one another in form
and color that the untrained eye would never know
them to be relatives. Even when male and female are
grown on the same flower-stalk and enclosed in a
common perianth, the hermaphroditism is generally
but apparent, owing to the physiological barriers of
heteromorphism and dichogamy. Sex-separation, indeed, is not only distinct among flowering plants, but
is kept up by a variety of complicated devices, and a
return to hermaphroditism is prevented by the most
elaborate precautions.

When we turn to the animal kingdom again, the
same great contrast arrests us. Half a century ago,
when Balbiani described the male and female elements
in microscopic infusorians, his facts were all but
rejected by science. But further research has placed
it beyond all doubt that the beginnings of sex are
synchronous almost with those shadowings in of life.
From a state marked by a mere varying of the nuclear
elements, a state which might almost be described as
one antecedent to sex, the sex-distinction slowly
gathers definition, and passing through an infinite
variety of forms, and with countless shades of
emphasis, reaches at last the climax of separateness
which is observed among birds and mammals. Often,
even in the Metazoa, this separateness is outwardly
obscured, as in star-fishes and reptiles ; often it is
matter of common observation ; while sometimes it is
carried to such a pitch of specialization that only the
naturalist identifies the two wholly unlike creatures
as male and female. Through the whole wide field
of Nature then this gulf is fixed. Each page of the
million-leaved Book of Species must be as it were split

in two, the one side for the male, the other for the female. Classification naturally takes little note of this distinction; but it is fundamental. Unlikenesses between like things are more significant than unlikenesses of unlike things. And the unlikenesses between male and female are never small, and almost always great. Though the fundamental difference is internal the external form varies; size, color, and a multitude of more or less striking secondary sexual characteristics separate the one from the other. Besides this, and more important than all, the cycle of a year's life is never the same for the male as for the female; they are destined from the beginning to pursue different paths, to live for different ends.

Now what does all this mean? To say that the sex-distinction is necessary to sustain the existence of life in the world is no answer, since it is at least possible that life could have been kept up without it. From the facts of Parthenogenesis, illustrated in bees and termites, it is now certain that Reproduction can be effected without fertilization; and the circumstance that fertilization is nevertheless the rule, proves this method of Reproduction, though not a necessity, to be in some way beneficial to life. It is important to notice this absence of necessity for sex having been created—the absence of any known necessity—from the merely physiological stand-point. Is it inconceivable that Nature should sometimes do things with an ulterior object, an ethical one, for instance? To no one with any acquaintance with Nature's ways will it be possible to conceive of such a purpose as the sole purpose. In these early days when sex was instituted it was a physical universe.

Undoubtedly sex then had physiological advantages; but when in a later day the ethical advantages become visible, and rise to such significance that the higher world nearly wholly rests upon them, we are entitled, as viewing the world from that higher level, to have our own suspicions as to a deeper motive underlying the physical.

Apart from bare necessity, it is further remarkable that no very clear advantage of the sex-distinction has yet been made out by Science. Hensen and Van Beneden are able to see in conjugation no more than a *Verjüngung* or rejuvenescence of the species. The living machinery in its wearing activities runs down and has to be wound up again; to keep life going some fresh impulse must be introduced from time to time; or the protoplasm, exhausting itself, seeks restoration in fertilization and starts afresh.[1] To Hatschek it is a remedy against the action of injurious variations; while to Weismann it is simply the source of variations. " I do not know," says the latter, " what meaning can be attributed to sexual reproduction other than the creation of hereditary individual characters to form the material on which natural selection may work. Sexual reproduction is so universal in all classes of multicellular organisms, and nature deviates so rarely from it, that it must necessarily be of pre-eminent importance. If it be true that new species are produced by processes of selection, it follows that the development of the whole organic world depends on these processes, and the part that amphigony has to play in nature, by rendering selection possible among multicellular organisms, is not only

[1] Geddes and Thomson, *The Evolution of Sex*, p. 163.

important, but of the very highest imaginable importance." [1]

These views may be each true; and probably, in a
measure, are; but the fact remains that the later
psychical implications of sex are of such transcendent
character as to throw all physical considerations into
the shade. When we turn to these, their significance
is as obvious as in the other case it was obscure.
This will appear if we take even the most dis-
tinctively biological of these theories—that of Weismann. Sex, to him, is the great source of variation in
Nature, in plainer English, of the variety of organisms
in the world. Now this variety, though not the main
object of sex, is precisely what it was essential for
Evolution by some means to bring about. The first
work of Evolution always is, as we have seen, to
create a mass of similar things—atoms, cells, men—
and the second is to break up that mass into as many
different kinds of things as possible. Aggregation
masses the raw material, collects the clay for the potter; differentiation destroys the featureless monotonies
as fast as they are formed, and gives them back in
new and varied forms. Now if Evolution designed,
among other things, to undertake the differentiation of
Mankind, it could not have done it more effectively
than through the device of sex. To the blending, or to
the mosaics, of the different characteristics of father
and mother, and of many previous fathers and
mothers, under the subtle wand of heredity, all the
varied interests of the human world is due. When
one considers the passing on, not so much of minute
details of character and disposition, but of the domi-

[1] *Biological Memoirs,* p. 281.

nant temperament and type, the new proportion in which already inextricably mingled tendencies are re-arranged, and the changed environment in which, with each new generation, they must unfold; it is seen how perfect an instrument for variegating humanity lies here. Had sex done nothing more than make an interesting world, the debt of Evolution to Reproduction had been incalculable.

THE ETHICAL SIGNIFICANCE OF MATERNITY.

But let us not be diverted from the main stream by these secondary results of the sex-distinction. A far more important implication lies before us. The problem that remains for us to settle is as to how the merely physical forms of Other-ism began to be accompanied or overlaid by ethical characters. And the solution of this problem requires nothing more than a consideration of the broad and fundamental fact of sex itself. In what it is, and in what it necessarily implies, we shall find the clue to the beginnings of the social and moral order of the world. For, rising on the one hand out of maleness and on the other hand out of femaleness, developments take place of such a kind as to constitute this the turning-point of the world's moral history. Let it be said at once that these developments are not to be sought for in the direction in which, from the nature of the factors, one might hastily suppose that they lay. What seems to be imminent at this stage, and as the natural end to which all has led up, is the institution of affection in definite forms between male and female. But we are on a very different track. Affection between male and female is a later, less fundamental, and, in its beginnings, less essential growth; and long prior to its existence, and largely the condition of it, is the even more beautiful development whose progress we have now to trace. The basis of this new development is indeed far removed from the mutual relations of sex with sex. For it lies in maleness and femaleness themselves, in their inmost quality and essential nature, in what they lead to and what they become. The superstructure, certainly, owes much to the psychical relations of father and mother, husband and wife, but the Evolution of Love began ages before these were established

What exactly maleness is, and what femaleness, has been one of the problems of the world. At least five hundred theories of their origin are already in the field, but the solution seems to have baffled every approach. Sex has remained almost to the present hour an ultimate mystery of creation, and men seem to know as little what it is as whence it came. But among the last words of modern science there are one or two which spell out a partial clue to both of these mysterious problems. The method by which this has been reached is almost for the first time a purely biological one, and if its inferences are still uncertain, it has at least established some important facts.

Starting with the function of nutrition as the nearest ally of Reproduction, the newer experimenters have discovered cases in which sex apparently has been determined by the quantity and quality of the food-supply. And in actual practice it has been found possible, in the case of certain organisms, to produce either maleness or femaleness by simply varying their nutrition—femaleness being an accompaniment of abundant food, maleness of the reverse. When Yung, to take an authentic experiment, began his observations on tadpoles, he ascertained that in the ordinary natural condition the number of males and females produced was not far from equal—the percentage being about 57 female to 43 male, thus giving the females a preponderance of seven. But when a brood of tadpoles was sumptuously fed the percentage of females rose to 78, and when a second brood was treated even more liberally, the number amounted to 81. In a third experiment with a still more highly nutritious diet, the result of the high feeding was more remarkable, for in this case 92 females were produced and only 8 males. In the case of butterflies and moths, it has been found that if caterpillars are starved before entering the chrysalis state the offspring are males, while others of the same brood, when highly nourished, develop into females. A still more instructive case is that of the aphides, the familiar plant-lice of our gardens. During the warmth of summer, when food is abundant, these insects produce parthenogenetically nothing but females, while in the famines of later autumn they give birth to males. In striking confirmation of this fact it has been proved that in a conservatory where the aphides enjoy perpetual summer, the parthenogenetic succession of females continued to go on for four years and stopped only when the temperature was lowered and food diminished. Then males were at once produced.[1] It will no longer be said that science is making no progress with this unique problem when it is apparently able to deter-

[1] *The Evolution of Sex*, pp. 41–46.

mine sex by turning off or on the steam in a green-house. With regard to bees the relation between nutrition and sex seems equally established. " The three kinds of inmates in a bee-hive are known to every one as queens, workers, and drones; or, as fertile females, imperfect females, and males. What are the factors determining the differences between these three forms? In the first place, it is believed that the eggs which give rise to drones are not fertilized, while those that develop into queens and workers have the normal history. But what fate rules the destiny of the two latter, determining whether a given ovum will turn out the possible mother of a new generation, or remain at the lower level of a non-fertile working female? It seems certain that the fate mainly lies in the quantity and quality of the food. Royal diet, and plenty of it, develops the future queens. . . . Up to a certain point the nurse bees can determine the future destiny of their charge by changing the diet, and this in some cases is certainly done. If a larva on the way to become a worker receive by chance some crumbs from the royal superfluity, the reproductive function may develop, and what are called 'fertile workers,' to a certain degree above the average abortiveness, result; or, by direct intention, a worker grub may be reared into a queen bee." [1]

It is unnecessary to prolong the illustration, for the point it is wished to emphasize is all but in sight. As we have just witnessed, the tendency of abundant nutrition is to produce females, while defective nutritive conditions produce males. This means that in so far as nutrition re-acts on the bodies of animals—and

[1] *TheEvolution of Sex*, p. 42. See also pp. 41–46.

nothing does so more—there will be a growing difference, as time begins to accumulate the effects, between the organization and life-habit of male and female respectively. In the male, destructive processes, a preponderance of waste over repair, will prevail; the result will be a katabolic habit of body; in the female the constructive processes will be in the ascendant, occasioning an opposite or anabolic habit. Translated into less technical language, this means that the predominating note in the male will be energy, motion, activity; while passivity, gentleness, repose, will characterize the female. These words, let it be noticed, psychical though they seem, are yet here the coinages of physiology. No other terms indeed would describe the difference. Thus Geddes and Thomson: " The female cochineal insect, laden with reserve-products in the form of the well-known pigment, spends much of its life like a mere quiescent gall on the cactus plant. The male, on the other hand, in his adult state, is agile, restless, and short-lived. Now this is no mere curiosity of the entomologist, but in reality a vivid emblem of what is an average truth throughout the world of animals—the preponderating passivity of the females, the freedomness and activity of the males." Rolph's words, because he writes neither of men nor of animals, but goes back to the furthest recess of Nature and characterizes the cell itself, are still more significant : " The less nutritive and therefore smaller, hungrier, and more mobile organism is the male ; the more nutritive and usually more quiescent is the female."

Now what do these facts indicate? They indicate that maleness is one thing and femaleness another, and that each has been specialized from the beginning to play a separate rôle in the drama of life. Among primitive peoples, as largely in modern times, "The tasks which demand a powerful development of muscle and bone, and the resulting capacity for intermittent spurts of energy, involving corresponding periods of rest, fall to the man ; the care of the children and all the various industries which radiate from the hearth, and which call for an expenditure of energy more continuous, but at a lower tension, fall to the woman." [1] Whether this or any theory of the origin of Sex be proved or unproved, the fact remains, and is everywhere emphasized in Nature, that a certain constitutional difference exists between male and female, a difference inclining the one to a robuster life and implanting in the other a certain mysterious bias in the direction of what one can only call the womanly disposition.

On one side of the great line of cleavage have grown up men—those whose lives for generations and generations have been busied with one particular set of occupations; on the other side have lived and developed women—those who for generations have been busied with another and a widely different set of occupations. And as occupations have inevitable reactions upon mind, character, and disposition, these two have slowly become different in mind and character and disposition. That cleavage, therefore, which began in the merely physical region, is now seen to extend into the psychical realm, and ends by supplying the world with two great and forever separate types. No efforts, or explanations, or expostulations

[1] Havelock Ellis, *Man and Woman*, p. 2.

can ever break down that distinction between maleness and femaleness, or make it possible to believe
that they were not destined from the first of time to
play a different part in human history. Male and
female never have been and never will be the same.
They are different in origin; they have travelled to
their destinations by different routes; they have had
different ends in view. The result is that they are
different, and the contribution therefore of each to the
evolution of the human race is special and unique.
By and bye it will be our duty to mark what Man, in
virtue of his peculiar gift, has done for the world;
part indeed of his contribution has been already re-
corded here. To him has been mainly assigned the
fulfilment of the first great function—the Struggle for
Life. Woman, whose higher contribution has not yet
been named, is the chosen instrument for carrying on
the Struggle for the Life of Others. Man's life, on
the whole, is determined chiefly by the function of
Nutrition; Woman's by the function of Reproduction.
Man satisfies the one by going out into the world, and
in the rivalries of war and the ardors of the chase, in
conflict with Nature, and amid the stress of industrial
pursuits, fulfilling the law of Self-preservation;
Woman completes her destiny by occupying herself
with the industries and sanctities of the home, and
paying the debt of Motherhood to her race.

Now out of this initial difference—so slight at first
as to amount to no more than a scarcely perceptible
bias—have sprung the most momentous issues. For
by every detail of their separate careers the two
original tendencies—to outward activity in the man;
to inward activity, miscalled passivity, in the woman

17

—became accentuated as time went on. The one life tended towards selfishness, the other towards unself-ishness. While one kept Individualism alive, the other kept Altruism alive. Blended in the children, these two master-principles from this their starting-point acted and re-acted all through history, seeking that mean in which true life lies. Thus by a Division of Labor appointed by the will of Nature, the conditions for the Ascent of Man were laid.

But by far the most vital point remains. For we have next to observe how this bears directly on the theme we set out to explore—the Evolution of Love. The passage from mere Other-ism, in the physiological sense, to Altruism in the moral sense, occurs in connection with the due performance of her natural task by her to whom the Struggle for the Life of Others is assigned. That task, translated into one great word, is Maternity—which is nothing but the Struggle for the Life of Others transfigured, transferred to the moral sphere. Focused in a single human being, this function, as we rise in history, slowly begins to be accompanied by those heaven-born psychical states which transform the femaleness of the older order into the Motherhood of the new. When one follows Maternity out of the depths of lower Nature, and beholds it ripening in quality as it reaches the human sphere, its character, and the character of the processes by which it is evolved, appear in their full divinity. For of what is Maternity the mother? Of children? No; for these are the mere vehicle of its spiritual manifestation. Of affection between female and male? No; for that, contrary to accepted beliefs, has little to do in the first instance with sex-relations. Of what then? Of Love itself, of Love as Love, of Love as Life, of Love as Humanity, of Love as the pure and undefiled fountain of all that is eternal in the world. In the long stillness which follows the crisis of Maternity, witnessed only by the new and helpless life which is at once the last expression of the older function and the unconscious vehicle of the new, Humanity is born. By an alchemy which remains, and must ever remain, the secret of Nature, the physiological forces give place to those higher principles of sympathy, solicitude, and affection which from this time onwards are to change the course of Evolution and determine a diviner destiny for a Human Race:

> "Earth's insufficiency
> Here grows to event;
> The indescribable
> Here it is done;
> The woman-soul leadeth us
> Upward and on." [1]

So stupendous is this transition that the mere possibility staggers us. Separated by the whole diameter of conscious intelligence and will, what possible affinities can exist between the Reproductive and the Altruistic process? What analogy can ever exist between the earlier physiological Struggle for the Life of Others and the later Struggle of Love? Yet, different though their accompaniments may be, when closely examined they are seen, at every essential point, running parallel with each other. The object in either case is to continue the life of the Species;

[1] *Faust*, Pt. II. Bayard Taylor's tr.

the essence of both is self-sacrifice; the first manifestation of the sacrifice is to make provision for Others
by helping them to draw the first few breaths of life.
But what has Love to do with Species? Can Altruism have reference to mere life? The answer is, that
in its first beginnings it has almost nothing to do
with anything else. For, consider the situation. Reproduction, let us suppose, has done its most perfect
work on the physiological plane: the result is that a
human child is born into the world. But the work of
Reproduction being to Struggle for the Life of the
Species, its task is only complete when it secures that
the child, representing the Species, shall live. If the
child dies, Reproduction has failed; the Species, so
far as this effort is concerned, comes to an end. Now,
can Reproduction as a merely physiological function
complete this process? It cannot. What can?
Only the Mother's Care and Love. Without these,
in a few hours or days, the new life must perish; the
earlier achievement of Reproduction is in vain.
Hence there comes a moment when these two functions meet, when they act as complements to each
other; when Physiology hands over its unfinished
task to Ethics; when Evolution—if for once one may
use a false distinction—depends upon the "moral"
process to complete the work the "cosmic" process
has begun.

At what precise stage of the Ascent, in association
with what class of animals, Other-ism began to shade
into Altruism in the ethical sense, is immaterial.
Whether the Altruism in the early stages is real or
apparent, profound or superficial, voluntary or automatic, does not concern us. What concerns us is that

the Altruism is there; that the day came when, even though a rudiment, it was a reality; above all that the arrangements for introducing and perfecting it were realities. The prototype, for ages, may have extended only to form, to the outward relation; for further ages no more Altruism may have existed than was absolutely necessary to the preservation of the Species. But to fix the eye upon it at that remote stage and assert that, because it was apparently then automatic, it must therefore have been automatic ever after, is to forget the progressive character of Evolution as well as to ignore facts. While many of the apparent Other-regarding acts among animals are purely selfish and purely automatic, undoubtedly there are instances where more is involved. Apart from their own offspring—in relation to which there may always be the suspicion of automatism; and apart from domestic animals—which are open to the further suspicion of having been trained to it—animals act spontaneously towards other animals; they have their playmates; they make friendships and very attached friendships. Much more, indeed, has been claimed for them; but it is not necessary to claim even this much. No evolutionist would expect among animals—domestic animals always excepted—any considerable development of Altruism, because the physiological and psychical conditions which directly led to its development in Man's case were fulfilled in no other creature.[1]

[1] The answer to the argument in favor of automatism is thus summarized by C. M. Williams: "(1) That functions which are preserved and inherited must evidently be, not only in animals and plants, but also and equally in man, such as favor the preser-

Simple as seems the method by which the first few sparks of Love were nursed into flame in the bosom of Maternity, the details of the evolution are so intricate as to require a chapter to themselves. But the emphasis which Nature puts on this process may be judged of by the fact that one-half the human race had to be set apart to sustain and perfect it. To the evolutionist who discerns the true proportions of the forces which made for the Ascent of Man, one of the two or three great events in the natural history of the world was the institution of sex. It is here that the master-forces which were to dominate the latest and highest stages of the process start; here, specialized into Egoism and Altruism, they part; and here, each having run its different course, they meet to distribute their gains to a succeeding race. With the initial impulses of their sex strengthened by the different life-routine to which each led, these two forces ran their course through history, determining by their ceaseless reactions the order and progress of

vation of the species ; those which do not so favor it must perish with the individuals or species to which they belong ; (2) that it cannot, indeed, be assumed that a result which has never come within the experience of the species can be willed as an end, although, with the species, function securing results which, from a human point of view, might be regarded as such, may be preserved; but (3) that, as far as we assume the existence of consciousness at all in any species or individual, we must assume pleasure and pain, pleasure in customary function, pain in its hindrance; and (4) that, as far as we can assume memory, we may al o feel authorized to assume that a remembered action may be associated with remembered results that come within the experience of the animal, some phases of which may thus become, as combined with pleasure or pain, ends to seek or consequences to avoid."—*Evolutional Ethics*, p. 386.

the world, or when wrongly balanced, its disorder and decay. According to evolutional philosophy there are three great marks or necessities of all true development—Aggregation, or the massing of things; Differentiation, or the varying of things; and Integration, or the re-uniting of things into higher wholes. All these processes are brought about by sex more perfectly than by any other factor known. From a careful study of this one phenomenon, science could almost decide that Progress was the object of Nature, and that Altruism was the object of Progress.

This vital relation between Altruism in its early stages and physiological ends, neither implies that it is to be limited by these ends nor defined in terms of them. Everything must begin somewhere. And there is no aphorism which the labors of Evolution, at each fresh beginning, have tended more consistently to endorse than "first that which is natural, then that which is spiritual." How this great saying also disposes of the difficulty, which appears and reappears with every forward step in Evolution, as to the qualitative terms in which higher developments are to be judged, is plain. Because the spiritual to our vision emerges from the natural, or, to speak more accurately, is convoyed upwards by the natural for the first stretches of its ascent, it is not necessarily contained in that natural, nor is it to be defined in terms of it. What comes "first" is not the criterion of what comes last. Few things are more forgotten in criticism of Evolution than that the nature of a thing is not dependent on its origin, that one's whole view of a long, growing, and culminating process is not to be governed by the first sight the microscope can catch of it. The processes of Evolution evolve as well as the products; evolve with the products. In the Environments they help to create, or to make available, they find a field for new creations as well as further reinforcements for themselves. With the creation of human children Altruism found an area for its own expansion such as had never before existed in the world. In this new soil it grew from more to more, and reached a potentiality which enabled it to burst the trammels of physical conditions, and overflow the world as a moral force. The mere fact that the first uses of Love were physical shows how perfectly this process bears the stamp of Evolution. The later function is seen to relieve the earlier at the moment when it would break down without it, and continue the ascent without a pause.

If it be hinted that Nature has succeeded in continuing the Ascent of Life in Animals without any reinforcement from psychical principles, the first answer is that owing to physiological conditions this would not have been possible in the case of Man. But even among animals it is not true that Reproduction completes its work apart from higher principles, for even there, there are accompaniments, continually increasing in definiteness, which at least represent the instincts and emotions of Man. It is no doubt true that in animals the affections are less voluntarily directed than in the case of a human mother. But in either case they must have been involuntary at first. It can only have been at a late stage in Evolution that Nature could trust even her highest product to carry on the process by herself. Before Altruism was strong enough to take its own initiative, necessity had to be laid upon all mothers, animal and human, to act in the way required. In part physiological, this necessity was brought about under the ordinary action of that principle which had to take charge of everything in Nature until the will of Man appeared—Natural Selection. A mother who did not care for her children would have feeble and sickly children. Their children's children would be feeble and sickly children. And the day of reckoning would come when they would be driven off the field by a hardier, that is a better-mothered, race. Hence the premium of Nature upon better mothers. Hence the elimination of all the reproductive failures, of all the mothers who fell short of completing the process to the last detail. And hence, by the law of the Survival of the Fittest, Altruism, which at this stage means good-motherism, is forced upon the world

This consummation reached, the foundations of the human world are finished. Nothing foreign remains to be added. All that need happen henceforth is that the Struggle for the Life of Others should work out its destiny. To follow out the gains of Reproduction from this point would be to write the story of the nations, the history of civilization, the progress of Social Evolution. The key to all these processes is here. There is no intelligible account of the world which is not founded on the realization of the place of this factor in development. Sociology, practically, can only beat the air, can make no step forward as a science, until it recognizes this basis in biology. It is the failure, not so much to recognize the supremacy of this second factor, but to see that there is any second factor at all, that has vitiated almost every attempt to construct a symmetrical social philosophy. It has long, indeed, been perceived that society is an organism, and an organism which has grown by natural growth like a tree. But the tree to which it is usually likened is such a tree as never grew on this earth. For it is a tree without flowers; a tree with nothing but stem and leaves; a tree that performed the function of Nutrition, and forgot all about Reproduction. The great unrecognized truth of social science is that the Social Organism has grown and flowered and fruited in virtue of the continuous activities and inter-relations of the two co-related functions of Nutrition and Reproduction, that these two dominants being at work it could not but grow, and grow in the way it has grown. When the dual nature of the evolving forces is perceived; when their reactions upon one another are understood; when the changed material with which they have to work from time to time, the further obstacles confronting them at every stage, the new Environments which modify their action as the centuries add their growths and disencumber them of their withered leaves,—when all this is observed, the whole social order falls into line. From the dawn of life these two forces have acted together, one continually separating, the other continually uniting; one continually looking to its own things, the other to the things of Others. Both are great in Nature—but "the greatest of these is Love."

CHAPTER VIII.

THE EVOLUTION OF A MOTHER.

THE Evolution of a Mother, in spite of its half-humorous, half-sacrilegious sound, is a serious study in Biology. Even on its physical side this was the most stupendous task Evolution ever undertook. It began when the first bud burst from the first plant-cell, and was only completed when the last and most elaborately wrought pinnacle of the temple of Nature crowned the animal creation.

What was that pinnacle? There is no more instructive question in science. For the answer brings into relief one of the expression-points of Nature—one of these great teleological notes of which the natural order is so full, and of which this is by far the most impressive. Run the eye for a moment up the scale of animal life. At the bottom are the first animals, the Protozoa. The Cœlenterata follow, then in mixed array, the Echinoderms, Worms, and Molluscs. Above these come the Pisces, then the Amphibia, then the Reptilia, then the Aves, then—What? The Mammalia, THE MOTHERS. There the series stops. Nature has never made anything since.

Is it too much to say that the one motive of organic

Nature was to make Mothers? It is at least certain that this was the chief thing she did. Ask the Zoologist what, judging from science alone, Nature aspired to from the first, he could but answer Mammalia—Mothers. In as real a sense as a factory is meant to turn out locomotives or clocks, the machinery of Nature is designed in the last resort to turn out Mothers. You will find Mothers in lower nature at every stage of imperfection; you will see attempts being made to get at better types; you find old ideas abandoned and higher models coming to the front. And when you get to the top you find the last great act was but to present to the world a physiologically perfect type. It is a fact which no human Mother can regard without awe, which no man can realize without a new reverence for woman and a new belief in the higher meaning of Nature, that the goal of the whole plant and animal kingdoms seems to have been the creation of a family, which the very naturalist has had to call Mammalia.

That care for others, from which the Mammalia take their name, though reaching its highest expression there, is introduced into Nature in cruder forms almost from the dawn of life. In the vegetable kingdom, from the motherlessness of the early Cryptogams, we rise to find a first maternity foreshadowed in the flowering tree. It elaborates a seed or nut or fruit with infinite precaution, surrounding the embryo with coat after coat of protective substance, and storing around it the richest foods for its future use. And rudimentary though the manifestation be, when we remember that this is not an incident in the tree's life but its whole blossom and crown, it is impossible but to think of this solicitude and Motherhood to-gether. So exalted in the tree's life is this provision for others that the Botanist, like the Zoologist, places the mothering plants at the top of his department of Nature. His highest division is the *Phanerogams* —named, literally, in terms of their reproductive specialization.

Crossing into the animal kingdom we observe the same motherless beginning, the same cared-for end. All elementary animals are orphans; they know neither home nor care; the earth is their only mother or the inhospitable sea; they waken to isolation, to apathy, to the attentions only of those who seek their doom. But as we draw nearer the apex of the animal kingdom, the spectacle of a protective Maternity looms into view. At what precise point it begins it is difficult to say. But that it does not begin at once, that there is a long and gradual Evolution of Mater-nity, is clear. From casual observation, and from pop-ular books, it might be inferred that care of offspring —we cannot yet speak of affection—is characteristic of the whole field of Nature. On the contrary, it is doubtful whether in the Invertebrate half of Nature it exists at all. If it does it is very rare; and in the Vertebrates it is met with only exceptionally till we reach the two highest classes. What does exist, and sometimes in marvellous perfection, is care for *eggs ;* but that is a wholly different thing, both in its phys-ical and psychical aspect, from love of offspring. The truth is Nature so made animals in the early days that they did not need Mothers. The moment they were born they looked after themselves, and were per-fectly able to look after themselves. Mothers in these days would have been a superfluity. All that Nature worked at at that dawning date was Maternity in a physical sense—Motherhood came as a later and a rarer growth. The children of those days were not really children at all ; they were only offspring, springers off, deserters from home. At one bound they were out into life on their own account, and she who begat them knew them no more. That early world, therefore, for millions and millions of years was a bleak and loveless world. It was a world without children and a world without Mothers. It is good to realize how heartless Nature was till these arrived.

In the lower reaches of Nature, things remain still unchanged. The rule is not that the Mother ignores, but that she never sees her child. The land-crabs of the West Indies descend from their homes in the mountains once a year, march in procession to the sea, commit their eggs to the waves, and come away. The burying-beetles deposit their fragile capsules in the dead carcase of a mouse or bird, plant all together in the earth, and leave them to their fate. Myriads of other creatures are born into the world, and ordained so to be born, whose Mothers are dead before they begin to live. The moment of birth with the Ephemeridæ is also the moment of death. These are not cases nevertheless where there has been no care. On the contrary, there is a solicitude for the egg of the most extreme kind—for its being placed exactly in the right spot, at the right time, protected from the weather, shielded from enemies, and provided with a first supply of food. The butterfly places the eggs of its young on the very leaf which the coming cater-pillar likes the most, and on the under side of the leaf where they will be least exposed—a case which illustrates in a palpable way the essential difference between Motherhood and Maternity. Maternity here, in the restricted sense of merely adequate physical care, is carried to its utmost perfection. Everything that can be done for the egg is done. Motherhood, on the other hand, is non-existent, is even an anatomical impossibility. If a butterfly could live till its egg was hatched—which does not happen—it would see no butterfly come out of the egg, no airy likeness of itself, but an earth-bound caterpillar. . If it recognized this creature as its child, it could never play the Mother to it. The anatomical form is so different that were it starving it could not feed it, were it threatened it could not save it, nor is it possible to see any direction in which it could be of the slightest use to it. It is obvious that Nature never intended to make a Mother here; that all that she desired as yet was to perfect the first maternal instinct. And the tragedy of the situation is that on that day when her training to be a true Mother should begin, she passes out of the world.

But there is another reason, in addition to the precocity of the offspring, why parental care is a drug in the market in lower Nature. There are such multitudes of these creatures that it is scarcely worth caring for them. The humbler denizens of the world produce offspring, not by units or tens, but by thousands and millions; and with populations so vast, maternal protection is not required to sustain the existence of the species. It was probably on the whole a better arrangement to produce a million and let them take their chance, than to produce one and take special

trouble with it. It was easier, moreover, a thousand
times easier, for Nature to make a million young than
one Mother. But the ethical effect, if one may use .
such a term here, of this early arrangement was nil.
All this saving of Motherly trouble meant for a long
space in Nature complete absence of maternal training. With children of this sort, Motherhood had no
chance. There was no time to love, no opportunity to
love, and no object to love. It was a period of physical installations; and of psychical installations only as
establishing the first stages of the maternal instinct
—the prenatal care of the egg. This is a necessary
beginning, but it is imperfect ; it arrests itself at the
critical point—where care can react upon the Mother.

Now, before Maternal Love can be evolved out of
this first care, before Love can be made a necessity,
and carried past the unhatched egg to the living thing
which is to come out of it, Nature must alter all her
ways. Four great changes at least must be introduced
into her programme. In the first place, she must
cause fewer young to be produced at a birth. In the
second place, she must have these young produced in
such outward form that their Mothers will recognize
them. In the third place, instead of producing them
in such physical perfection that they are able to go
out into life the moment they are born, she must
make them helpless, so that for a time they must
dwell with her if they are to live at all. And fourth-
ly, it is required that she shall be made to dwell with
them ; that in some way they also should be made
necessary—physically necessary—to her to compel her
to attend to them. All these beautiful arrangements
we find carried out to the last detail. A mother is

made, as it were, in four processes. She requires, like the making of a colored picture, four separate paint· ings, each adding some new thing to the effect. Let us note the way in which woman—savage woman— became caretaker, and watcher, and nurse, and passed from femaleness to the higher heights of Mother· hood.

The first great change that had to be introduced into Nature was the diminishing of the number of young produced at a birth. As we have seen, nearly all the lower animals produce scores, or hundreds, or thousands, or millions, at one time. Now, no mother can love a million. Clearly, if Nature wishes to make caretakers, she must moderate her demands. And so she sets to work to bring down the numbers, reducing them steadily until so few remain that Motherhood becomes a possibility. How great this change is can only be understood when one realizes the almost in- calculable fecundity of the first-created forms of life. When we examine the progeny of the lowest plants we find ourselves among figures so high that no mi- croscope can count them. The *Protococcus Nivalis* shows its exuberant reproductive power by reddening the Arctic landscape with its offspring in a single night. When we break or shake the Puff-ball of the well-known fungus, the cloud of progeny darkens the air with a smoke made up of uncountable millions of spores. *Hydatina Senta*, one of the Rotifera, propa- gates four times in thirty-four hours, and in twelve days is the parent of sixteen million young. Among fish the number is still very great. The herring and the cod give birth to a million ova, the frog spawns eggs by the thousand, and most of the creatures at
18

and below that level in a like degree. Then comes a
gradual change. When we pass on to the Reptiles,
the figures fall into hundreds. On reaching the birds
the young are to be counted by tens or units. In the
highest of Mammals the rule is one. This bringing
down of the numbers is a remarkable circumstance.
It means the calling in of a diffused care, to focus it
upon one, and concentrate it into Love.

The next thing was to make it possible for the parent to recognize its young. If it was difficult to love
a million it was impossible to love an embryo. In the
lower reaches the young are never in the smallest
degree like their parents, and, granting the highest
power of recognition to the Mother, it is impossible
that she should recognize her own offspring. For
generations even Science was imposed upon here, for
many forms of life were described and classified as
distinct species which have turned out to be simply
the young of other species. It may be useless to contrast so striking a case as the ciliated *Planula* with
the adult *Aurelia*—vagaries of form which for generations deceived the naturalist—for it is doubtful
whether creatures of the Medusoid type have eyes;
but in the higher groups, where power of recognition
is more certain, the unlikeness of progeny to parent is
often as decided. The larval forms of the Star fish,
or the Sea Urchin, or their kinsman the Holothurian
are disguised past all recognition; and among the
Insects the relation between Butterflies and Moths
and their respective caterpillars is beyond any possible
clue. No doubt there are other modes of recognition
in Nature than those which depend on the sense of
sight. But looked at on every side, the fact remains

that the power to identify their young is all but absent until the higher animals appear.

The next work of Nature, therefore, was to make the young resemble the parent, to make, in short, the children presentable at birth. And the means taken to effect this are worth noting. Nature always makes her changes with a marvellous economy, and generally, as in this case, with a quite startling simplicity. To start making a new kind of embryo, a plan obvious to us, was not thought of. That would have been to have lost all the time spent on them already. If Nature begins a thing and wishes to make a change, she never goes back to the beginning and starts *de novo*. Her respect for her own work is profound. To begin at the beginning again would not only be lost work, but waste of future time; and Evolution, slow as it may seem, never fails to take the quickest path. She did not then start making new embryos. She did not even touch up the old embryos. All that she did was to *keep them hidden* till they grew more presentable. She left them exactly as they were, only she drew a veil over them. Instead of saying " Let us re-create these little things," she passed the word " Let us delay them till they are fair to see." And from the day that word was passed, the embryos were hindered in the eggs, and the eggs were hindered in the nest, and the young were hindered in the body, retained in the dark for weeks and months, so that when first they caught the Mother's eye they were " strong and of a good liking."

Though in no case in higher Nature is the young an exact reproduction of its parent, it will be admitted that the likeness is very much greater than among any of the lower animals. The young of many birds are at least a colorable imitation of their parents; Nature's young geese are at least like enough geese not to be mistaken for swans; no dog could be misled into mistaking—even apart from the sense of smell—a kitten for a puppy, nor would a hare ever be taken in by the young of a rabbit. Among domestic animals like the sheep and cow there is a culmination of adaptation in this direction, the lamb and the calf when born being almost fac-similes of their Mothers. But this point need not be dwelt on. It is of insignificant importance, and belongs to the surface. The idea of Nature going out of her way to make better family likenesses will not stand scrutiny as a final end in physiology. These illustrations are simply adduced to confirm the impression that Nature is working not aimlessly, not even mysteriously, but in a specific direction; that somehow the idea of *Mothers* is in her mind, and that she is trying to draw closer and closer the bonds which are to unite the children of men. It will be enough if we have gathered from this parenthesis that some time in the remote past, parent and child came to be introduced to one another; that the young when born into the world gradually approached the parental form, that they no longer " shocked them by their larval ugliness "; so that " the first human mother on record, seeing her first-born son, exclaimed · ' I have gotten a *Man* from the Lord.' " [1]

If this second process in the Evolution of Motherhood is of minor importance, the necessity for the third will not be doubted. What use is there for perfecting the power of recognition between parent and

[1] *Mammalian Descent*, Prof. W. P. Parker, F. R. S., p. 14.

child if the latter act like the run of offspring in lower nature—spring off into independent life the moment they are born? If the Mother is to be taught to know her progeny, surely the progeny also must be taught not to abandon their Mother. And hence Nature had to set about a somewhat novel task—to teach the youth of the world the Fifth Commandment. Glance once more over the Animal series and see how thoroughly she taught them the lesson. It is sometimes said that Nature has no imperatives. In reality it is all imperative. This Commandment was thrust upon the early world under penalties for disobedience the most exacting that could be devised—-the threat of death. Pick out a few children and inspect them. Take one from the bottom of Nature, one from the middle, and one from the top, and see if any progress in filial duty is visible as we ascend. The first,—-the young of Aurelia will do, or a ciliated Infusorian,—representing countless millions like itself, is the Precocious Child. The moment this embryo is born it leaves the domestic hearth; the chances are it has never seen its parents. If it has it disowns them on the spot. A better swimmer in many cases—for many of the parents have forgotten how to swim—it cannot be overtaken. It ignores its Mother and despises her. The second is the Good Intentioned Child. This child —a bird, let us say—begins well, stays much at home in the early days, but plays the prodigal towards the close. For some weeks it remains quietly in the egg; for more weeks it remains—not quite so quietly—in the nest; and for more weeks still—but with an obvious itching to be off—in the neighborhood of the nest. This, nevertheless, is a good subject. It is

really a sort of a child, and has really a sort of a
Mother. The third is the Model Child—the Mammal.
In this child, which is only found in the high places of
Nature, infancy reaches its last perfection. Housed,
protected, sumptuously fed, the luxurious children
keep to their Mother's side for months and years, and
only quit the parental roof when their filial education
is complete.

On a casual view of the Examiner's Report on these
various children of Nature the physiologist, as distinguished from the educationalist, might object that
so far from being the subject of congratulation it is a
clear case for censure. If early Nature could turn out
ready-made animals in a single hour, is it not a retro-
grade move to have to take so long about it later on?
When one contrasts the free swimming embryo of a
Medusa, dashing out into its heroic life the moment it
is born, with the helpless kitten or the sightless pup,
is it unfair to ask if Nature has not lost the trick of
making lusty lives? Is she not trying the new experiment at the risk of blundering the old one, and why
cannot she continue the earlier and more brilliant
device of making her children knight-errants from the
first? Because brilliance is not her object. Her object is ethical as well as physiological; and though
when we look below the surface a purely physiological
explanation of the riddle will appear, the ethical gain
is not less clear. By curbing them she is educating
them, taming them, rescuing them from a wild and
lawless life. These roving embryos are mere bandits;
their nature and habits must be changed; not a
sterner race but a gentler race must be born. New
words must come into the world—Home, Love,

Mother. And these imperceptibly slow drawings together of parent and child are the inevitable preliminaries of the domestication of the Human Race. Regarded from the ethical point of view there are few
things more significant than this reining-in of the
world's rampant youth, this tightening the bonds of
family life, this most gentle introduction of gentleness
into a world cold with motherless children and heartless with childless mothers.

The personal tie once formed between parent and
offspring could never be undone, and from this
moment onwards must grow from more to more. For
observe what has happened. A generation has grown
up to whom this tie is the necessity of existence.
Every Mammalian child born into the world must
come to be fed, must, for a given number of hours
each day, be in the maternal school, and whether it
like it or not, learn its lessons. No young of any
Mammal can nourish itself. There is that in it therefore at this stage which compels it to seek its Mother;
and there is that in the Mother which compels it even
physically—and this is the fourth process, on which
it is needless to dwell—to seek her child. On the
physiological side, the name of this impelling power
is lactation; on the ethical side, it is Love. And
there is no escape henceforth from communion between Mother and child, or only one—death. Break
this new bond and the Mammalia become extinct.
Nature is in earnest here, if anywhere. The training
of Humanity is seen to be under a compulsory education act. It is in the severity and dread of her penalties, coupled with the impossibility of evading the
least of them, that the will of Nature and the serious-

ness of her purposes are most declared. For the
physiological gains which underlie these ethical relations are all-important. It is largely owing to them
that the Mammalia have taken their place in the van
of the procession of life. Under the earlier system
life had a bad start ; each animal had to push its way
upward single-handed from the egg. It was planted,
so to speak, on the first rung of the ladder, and as the
risks of life are immeasurably great in infancy, it had
all these risks to take. Under the new system it is
launched into the battle already nourished and strong,
and passed scatheless through the first vicissitudes of
youth. In the higher Mammalia, in virtue of the
possession by this group of a placenta in addition to
the ordinary Mammalian characteristics, the young
have a double chance of a successful start. The
development, in fact, of higher forms of life on the
earth has depended on the physical perfecting of
Mothers, and of the physiological ties which bind
them to their young. With the immense structural
advance of the Mammalia, an order of being was introduced into Nature whose continuity as an all but
immortal series could never be broken. Thus whatever moral relations underlie the extraordinary
physical characteristic of this highest class of animals,
there is the added guarantee that they can never be
destroyed.

With the physical programme carried out to the
last detail, the ethical drama opened. An early
result, partly of her sex, and partly of her passive
strain, is the founding through the instrumentality of
the first savage Mother of a new and a beautiful social
state—Domesticity. While Man, restless, eager,

hungry, is a wanderer on the earth, Woman makes a Home. And though this Home be but a platform of sticks and leaves, such as the gorilla builds on a tree, it becomes the first great school-room of the human race. For one day there appears in this roofless room that which is to teach the teachers of the world—a Little Child.

No greater day ever dawned for Evolution than this on which the first human child was born. For there entered then into the world the one thing wanting to complete the Ascent of Man—a tutor for the affections. It may be that a Mother teaches a Child, but in a far deeper sense it is the Child who teaches the Mother. Millions of millions of Mothers had lived in the world before this, but the higher affections were unborn. Tenderness, gentleness, unselfishness, love, care, self-sacrifice—these as yet were not, or were only in the bud. Maternity existed in humble forms, but not yet Motherhood. To create Motherhood and all that enshrines itself in that holy word required a human child. The creation of the Mammalia established two schools in the world—the two oldest and surest and best equipped schools of Ethics that have ever been in it—the one for the Child, who must now at least know its Mother, the other for the Mother, who must as certainly attend to her Child. The only thing that remains now is to secure that they shall both be kept in that school as long as it is possible to detain them. The next effort of Evolution, therefore —the fifth process as one might call it—is to lengthen out these school days, and give affection time to grow.

No animal except Man was permitted to have his

education thus prolonged. Many creatures were allowed to stay at school for a few days or weeks, but
to one only was given a curriculum complete enough
to accomplish its exalted end. Watch two of the
highest organisms during their earliest youth, and
observe the striking contrast in the time they are made
to remain at their Mother's side. The first is a human
infant; the second, born, let us suppose, on the same
day, is a baby monkey. In a few days or weeks the
baby monkey is almost able to leave its Mother. Already it can climb, and eat, and chatter like its
parents; and in a few weeks more the creature is as
independent of them as the winged-seed is of the
parent tree. Meantime, and for many months to come,
its little twin is unable to feed itself, or clothe itself,
or protect itself; it is a mere semi-unconscious chattel,
a sprawling ball of helplessness, the world's one type
of impotence. The body is there in all its parts, bone
for bone and muscle for muscle, like the other. But
somehow this body will not do its work. Something
as yet hangs fire. The body has eyes but they see not,
ears but they hear not, limbs but they walk not. This
body is a failure. Why does the human infant lie like
a log on the forest-bed while its nimble prototype
mocks it from the bough above? Why did that which
is not human step out into life so long before that
which is?

The question has been answered for us by Mr. John
Fiske, and the world here owes to him one of the most
beautiful contributions ever made to the Evolution of
Man. We know what this delay means ethically—it
was necessary for moral training that the human child
should have the longest possible time by its Mother's

side—but what determines it on the physical side? The thing that constitutes the difference between the baby monkey and the baby man is an extra piece of machinery which the last possesses and the first does not. It is this which is keeping back the baby man. What is that piece of machinery? A brain, a human brain. The child, nevertheless, is not using it. Why? Because it is not quite fitted up. Nature is working hard at it; but owing to its intricacy and delicacy the process requires much time, and till all is ready the babe must remain a *thing*. And why does the monkey brain get ready first? Because it is an easier machine to make. And why should it be easier to make? Because it is only required to do the life-work of an Animal; the other has to do the life-work of a Man. Mental Evolution, in fact, here steps in, and makes an unexpected contribution to the ethical development of the world.

An apparatus for controlling one of the lower animals can be turned out from the workshop of Nature sometimes in a day. The wheels are few, the works are simple, the connections require little time for adjustment or correction. Everything that a humble organism will do has been done a million times by its parents, and already the faculties have been carefully instructed by heredity and will automatically repeat the whole life and movement of their race. But when a Man is made, it is not an automaton that is made. This being will do new things, think new thoughts, originate new ways of life. His immediate ancestors have done the same, but done some of them so seldom, and others of them for so short a time, that heredity has failed to notice them. For half the life, therefore, that lies before the human offspring no storage of habit has been handed down from the past. Each descendant must carve a way through the world for itself, and learn to comport itself through all the varying incidents of life as best it can. Now the equipment for this is very complex. Into the infant's frame must be fitted not only the apparatus for automatic repetition of what its parents have done, but the apparatus for intelligent initiation; not only the machinery for carrying on the involuntary and reflex actions— involuntary and reflex because they have been done so often by its ancestors as to have become automatic—but for the voluntary and self-conscious life which will do new things, choose fresh alternatives, seek higher and more varied ends. The instrument which will attend to breathing even when we forget it; the apparatus which will make the heart beat even though we try to stop it; the self-acting spring which makes the eyelid close the moment it is threatened— these and a hundred others are old and well-tried inventions which, from ceaseless practice generation after generation, work perfectly in each new individual from the start. Nature therefore need waste no time at this late day on their improvement. But the higher brain is comparatively a new thing in the world. It has to undertake a vaster range of duties, often totally new orders of duties; it has to do things which its forerunners had not quite learned to do, or had not quite learned to do unthinkingly, and the inconceivably complex machinery requires time to settle to its work. The older brain-processes have been greatly accelerated even now, and appear in full activity at an early stage in the infant's life, but the newer and the higher are in perfect order only after a consider-able interval of adjustment and elaboration.

Now Infancy, physiologically considered, means the fitting up of this extra machinery within the brain; and according to its elaborateness will be the time required to perfect it. A sailing vessel may put to sea the moment the rigging is in; a steamer must wait for the engines. And the compensation to the steamer for the longer time in dock is discovered by and bye in its vastly greater usefulness, its power of varying its course at will, and in its superior safety in time of war or storm. For its greater after-usefulness also, its more varied career, its safer life, humanity has to pay tribute to Evolution by a delayed and helpless Infancy, a prolonged and critical constructive process. Childhood in its early stage is a series of installations and trials of the new machinery, a slow experimenting with powers and faculties so fresh that heredity in handing them down has been unable to accompany them with full directions as to their use.

The Brain of Man, to change the figure—if indeed any figure of that marvellous molecular structure can be attempted without seriously misleading—is an elevated table-land of stratified nervous matter, furrowed by deep and sinuous cañons, and traversed by a vast net-work of highways along which Thoughts pass to and fro. The old and often-repeated Thoughts, or mental processes, pass along beaten tracks; the newer Thoughts have less marked footpaths; the newer still are compelled to construct fresh Thought-routes for themselves. Gradually these become established thoroughfares; but in the increasing traffic and complexity of life, new paths in endless multitudes have to be

added, and bye lanes and loops between the older highways must be thrown into the system. The stations
upon these roads from which the travellers set out
are cells; the roads are transit fibres; the travellers
themselves are in physiological language nervous dis-
charges, in psychological language mental processes.
Each new mental process involves a new redistribu-
tion of nervous matter among the cells, a new travel-
ling of nervous discharge along one or many of the
transit fibres. Now in every new connection of ideas
multitudes of cells and even multitudes of groups of
cells may be concerned, so that should it happen that
a combination of these precise centres had never been
made before, it is obvious that no routes could possibly exist between them, and these must then and
there be prospected. Each new Thought is therefore
a pioneer, a road-maker, or road-chooser, through the
brain; and the exhaustless possibilities of continuous
development may be judged from the endlessness of
the possible combinations. In the oldest and most-
used brain there must always remain vast territories
still to be explored, and as it were civilized; and in all
men multitudes of possible connections continue to the
last unrealized. When it is remembered, indeed, that
the brain itself is very large, the largest mass of
nerve-matter in the organic world; when it is further
realized that each of the cells of which it is built up
measures only one tenth-thousandth of an inch in
diameter, that the transit fibres which connect them
are of altogether unimaginable fineness, the limitless-
ness of the powers of Thought and the inconceivable
complexity of these processes will begin to be understood.

Now it is owing to the necessity for having a certain number of the more useful routes established before the babe can be trusted from its Mother's side, that the delay of Infancy is required. And even after the child has begun to practise the art of living for itself, time has still to be granted for many purposes —for new route-making, for becoming familiar with established thoroughfares, for practising upon obstacles and gradients, for learning to perform the journeys quickly and without fatigue, for allowing acts repeated to accelerate and embody themselves as habits. In the savage-state, where the after-life is simple, the adjustments are made with comparative ease and speed; but as we rise in the scale of civilization the necessary period of Infancy lengthens step by step, until in the case of the most highly educated man, where adjustments must be made to a wide intellectual environment, the age of tutelage extends for almost a quarter of a century.

The use of all this to morals, the reactions especially upon the Mother, are too obvious to dwell on. Till the brain arrived, everything was too brief, too rapid for ethical achievements; animals were in a hurry to be born, children thirsted to be free. There was no helplessness to pity, no pain to relieve, no quiet hours, no watching; to the Mother, no moment of suspense — the most educative moment of all—when the spark of life in her little one burned low. Parents could be no use to their offspring physically, and the offspring could be no use to their parents psychically. The young required no Infancy; the old acquired no Sympathy. Even among the other Mammalia or the Birds the

Mother's chance was small. There, Infancy extends to a few days or weeks, yet is but an incident in a life preoccupied with sterner tasks. A lioness will bleed for her cub to-day, and in to-morrow's struggle for life contend with it to the death. A sheep knows its lamb only while it is a lamb. The affection in these cases, fierce enough while it lasts, is soon forgotten, and the traces it left in the brain are obliterated before they have furrowed into habit. Among the Carnivora it is instructive to observe that while the brief span of infancy admits of the Mother learning a little Love, the father, for want of even so brief a lesson, remains untouched, so wholly untouched indeed that the Mother has often to hide her offspring from him lest they be devoured. Love then had no chance till the Human Mother came. To her alone was given a curriculum prolonged enough to let her graduate in the school of the affections. Not for days or weeks, but for months, as the cry of her infant's helplessness went forth, she must stand between the flickering flame and death; and for years to come, until the budding intellect could take its own command, this Love dare not grow cold, or pause an hour in its unselfish ministry.

Begin at the beginning again and recall the fact of woman's passive strain. A tendency to passivity means, among other things, a capacity to sit still. Be it but for a minute or an hour does not matter; the point is that the faintest possible capacity is there. For this is the embryo of Patience and if much and long nursed a fully fledged Patience will come out of it. Supply next to this new virtue some definite object on which to practise, let us say a child. When this child is in trouble the Mother will observe the signs of pain. Its cry will awaken associations, and in some dull sense the Mother will feel with it. But "feeling with another" is the literal translation of the name of a second virtue—Sympathy. From feeling with it, the parent will sooner or later be led to do something to help it; then it will do more things to help it; finally it will be always helping it. Now, to care for things is to become Careful; to tend things is to become Tender. Here are four virtues—Patience, Sympathy, Carefulness, Tenderness—already dawning upon mankind.

On occasion Sympathy will be called out in unusual ways. Crises will occur—dangers, famines, sicknesses. At first the Mother will be unable to meet these extreme demands—her fund of Sympathy is too poor. She cannot take any exceptional trouble, or forget herself, or do anything very heroic. The child, unable to breast the danger alone, dies. It is well that this should be so. It is the severity and righteous justice of Nature—the tragedy of Iván Ivánovitch anticipated by Evolution. A Mother who has failed in helpfulness must leave no successor to perpetuate her unworthiness in posterity. Somewhere else, however, developing along similar lines, there is another fractionally better Mother. When the emergency occurs, she rises to the occasion. For one hour she transcends herself. That day a cubit is added to the moral stature of mankind; the first act of Self-Sacrifice is registered in favor of the human race. It may or may not be that the child will acquire its Mother's virtue. But unselfishness has scored; its child has proved itself fitter to survive than the child

19

of Selfishness. It does not follow that in all circumstances the nobler will be always victorious : but it has a great chance. A few score more of centuries, a few more millions of Mothers, and the germs of Patience, Carefulness, Tenderness, Sympathy, and Self-Sacrifice will have rooted themselves in Humanity.

See then what the Savage Mother and her Babe have brought into the world. When the first Mother awoke to her first tenderness and warmed her loneliness at her infant's love, when for a moment she forgot herself and thought upon its weakness or its pain, when by the most imperceptible act or sign or look of sympathy she expressed the unutterable impulse of her Motherhood, the touch of a new creative hand was felt upon the world. However short the earliest infancies, however feeble the sparks they fanned, however long heredity took to gather fuel enough for a steady flame, it is certain that once this fire began to warm the cold hearth of Nature and give humanity a heart, the most stupendous task of the past was accomplished. A softened pressure of an uncouth hand, a human gleam in an almost animal eye, an endearment in an inarticulate voice—feeble things enough. Yet in these faint awakenings lay the hope of the human race. "From of old we have heard the monition, 'Except ye be as babes ye cannot enter the kingdom of Heaven'; the latest science now shows us—though in a very different sense of the words— that unless we had been as babes, the ethical phenomena which give all its significance to the phrase 'Kingdom of Heaven' would have been non-existent for us. Without the circumstances of Infancy, we might have become formidable among animals through sheer force of sharp-wittedness. But except for these circumstances we should never have comprehended the meaning of such phrases as ' self-sacrifice ' or ' devotion.' The phenomena of social life would have been omitted from the history of the world, and with them the phenomena of ethics and religion." [1]

[1] Fiske, *Cosmic Philosophy* Vol. II., p. 363.

CHAPTER IX.

THE EVOLUTION OF A FATHER.

In last chapter we watched the beautiful experiment of Nature making Mothers. We saw how the young produced at one birth were gradually reduced in numbers until it was possible for affection to concentrate upon a single object; how that object was delayed in birth till it was a likeable and presentable thing; how it was tied to its mother's side by physical bonds, and hindered there for years to give time for the Mother's care to ripen into love. We saw, what was still more instructive, that Nature, when she had laid the train for perfecting these arrangements, gave up making any more animals; and that there were physiological reasons why this well-mothered class should survive beyond all others, and, by sheer physiological fitness, henceforth dominate the world.

But there was still a crowning task to accomplish. The world was now beginning to fill with Mothers, but there were no Fathers. During all this long process the Father has not even been named. Nothing that has been done has touched or concerned him almost in the least degree. He has gone his own way

lived outside all these changes ; and while Na.
succeeded in moulding a human Mother and a hu..
child, he still wanders in the forest a savage and
unblessed soul.

This time for him, nevertheless, is not lost. In
his own way he is also at school, and learning lessons
which will one day be equally needed by humanity.
The acquisitions of the manly life are as necessary
to human character as the virtues which gather their
sweetness by the cradle; and these robuster elements
—strength, courage, manliness, endurance, self-reli-
ance—could only have been secured away from
domestic cares. Apart from that, it was not necessary to put the Father through the same mill as the
Mother. Whatever the Mother gained would be
handed on to her boys as well as to her girls, and
with the law of heredity to square accounts, it was
unnecessary for each of the two great sides of humanity to make the same investments. By one acquiring
one set of virtues and the other another, the blend in
the end would be the richer ; and, without obliterating the eternal individualities of each, the measure
of completeness would be gained more quickly for the
race. Before heredity, however, could do its work
upon the Father a certain basis had to be laid. With
his original habits he would squander the hereditary
gains as fast as he received them, and unless some
change was brought about in his mode of life the old
wild blood in his veins would counteract the gentler
influence, and leave all the Mother's work in vain.
Hence Nature had to set about another long and difficult process—to make the savage Father a reformed
character.

The Evolution of a Father is not so beautiful a process as the Evolution of a Mother, but it was almost as formidable a problem to attack. As much depended on it, as we shall see, as the training of the mother; and though it began later it required the bringing about of one or two changes in Nature as novel as any that preceded it. When the work was begun, the Father was in a much worse plight, so far as training for family life was concerned, than the Mother. If Maternity was at a feeble level in the lower reaches of Nature, Paternity was non-existent. Among a few Invertebrates the male parent took a passing share in the care of the egg, but it is not until we are all but at the top that fatherly interest finds any real expression. Among the Birds, the parents unite together in most cases to build the nest, the Father doing the rough work of bringing in moss and twigs, while the more trusty Mother does the actual work. When the eggs are laid, the male parent also takes his turn at incubation; supplies food and protection; and lingers round the place of birth to defend the fledglings to the last. When we leave the Birds, however, and pass on to the Mammals, the Fathers are nearly all backsliders. Many are not only indifferent to their young, but hostile: and among the Carnivora the Mothers have frequently to hide their little ones in case the father eats them.

We have another and a more serious count against early Fatherhood. If the Love of Father for child was in this backward state, infinitely more grave was the condition of things between him and the Mother. Probably we have all taken it for granted that husbands and wives have always loved one another.

Evolution takes nothing for granted. The affection between husband and wife is, of all the immeasurable forms of Love, the most beautiful, the most lasting, and the most divine ; yet up to this time we have not been able even to record its existence. The finished results of Evolution appear so natural to us, looking back from this late day, that we continually ignore the difficulties it had to meet, and forget how every single step in progress from the lowest to the highest had to be carried at the bayonet's point. The most informed naturalist probably has never given Nature credit for a thousandth part of the work she has done, or has succeeded in presenting to his mind more than a surface outline of the gigantic series of problems she had to solve. In lower Nature, as a simple fact, male and female do not love one another; and in the lower reaches of Human Nature, husband and wife do not love one another. Among exceptional nations, for the last few hours of the world's history, husbands and wives have truly loved ; but for the vast' mass of Mankind, during the long ages which preceded historic times, conjugal love was probably all but unknown.

Now here is a very pretty problem for Evolution. She has at once to make good Husbands and good Fathers out of lawless savages. Unless this problem is solved the higher progress of the world is at an end. It is the mature opinion of every one who has thought upon the history of the world, that the thing of highest importance for all times and to all nations is Family Life. When the Family was instituted, and not till then, the higher Evolution of the world was secured. Hence the exceptional value of the Father's development. As the other half of the arch on which the whole higher world is built, his taming, his domestication, his moral discipline, are vital; and in the nature of things this was the next great operation undertaken by Evolution.

The first step in the transition was to relate him, definitely and permanently, to the Mother. And here a formidable initial obstacle had to be encountered. The apathy and estrangement between husband and wife in the animal world is radical and universal. There is almost no such thing there as married life. Marriage, in anthropology, is not a word for an occasion, but for a state; it is not, that is to say, a wedding, but a dwelling together throughout life of husband and wife. Now when Man emerged from the animal creation this institution of conjugal life had not been arrived at. Marriage like everything else has been slowly evolved, and until it was evolved, until they learned to dwell continually together, there was no chance for mutual love to spring up between male and female. In Nature the pairing season is usually but an incident. It lasts only a very short time, and during the rest of the year, with some exceptions the sexes remain apart. From the investigations of Westermarck,—who has lately contributed to sociology the most masterly account of the Evolution of Marriage we possess—it appears more than probable that the earliest progenitors of Man had also a pairing season, and that the young were born at a particular time of the year, and never at any other time. All the animals nearest to Man in Nature have such a season, and there are only a few known—the elephant for instance, and some of the whales—which

have none. Now the brevity of this period in the
father's case must have told against his developing
any real affection. If he is to run away a few days
after the young are born he will miss all the discipline of the home, and as this discipline is essential, as
this is the only way in which love can be acquired,
or inherited love developed, some method must be
adopted in his case to extend the period of home life
during which it can act.

Now let us see how this was done. The problem
being to give Love time, the solution was in some way
to alter the circumstances which confined the pairing
season to a specific date—to abolish, in fact, the
pairing season in the case of Man, and lengthen
out the time in which husband and wife should stay
together. And as this was actually the method
adopted, we have first to ask what these special
circumstances were. Why should animals have specific dates at all? The clue will be found if we examine
carefully what these dates are and the reasons Nature
has had for choosing them. Some wise principle
must underlie this, or it would not be the universal
rule it is. The pairing time with Birds, as every one
knows, occurs in the Spring. With Reptiles this is
also the case; but among Mammals each species has a
season peculiar to itself, every separate month being
selected by one or other, and invariably adhered to.
"The bat pairs in January and February; the wild
camel in the desert to the east of Lake Lob-nor, from
the middle of January nearly to the end of February;
the Canis Azaræ and the Indian bison in winter; the
weasel in March; the kulan from May to July; the
musk-ox at the end of August; the elk, in the Baltic

Provinces, at the end of August, and, in Asiatic Rus-
sia, in September or October; the wild Yak in Tibet
in September; the reindeer in Norway at the end of
September; the badger in October; the Capra pyre-
naica in November; the chamois, the musk-deer, and
the orongo-antelope in November and December; the
wolf, from the end of December to the middle of
February." [1] It might seem that no law governed
these various dates, but their very variety is the proof
of an underlying principle. For these dates show
that each animal in each particular country chooses
that time of the year to give birth to her young when
they will have the best chance of surviving—that is to
say, when the climate is mildest, food most abundant,
and the prospects of life on the whole most favor-
able. The dormouse thus brings forth its young in
August, when the nuts begin to ripen; and the young
deer sees the light just before the first grass shoots
into greenness. Because those born at this season
survived and those born out of it perished, by the
prolonged action of Natural Selection these dates in
time probably became engrained in the species, and
would only alter with climate itself.

But when Man's Evolution made a certain progress,
and when the Mother's care reached mature perfection, it was no longer imperative for children to be
born only when the sun was shining, and the fruits
grew ripe. The parents could now make provision
for any weather and for any dearth. They could give
their little ones clothes when nights grew cold; they
could build barns and granaries against times of
famine. In any climate, and at any time, their young

[1] Westermarck's *History of Human Marriage*, p. 26.

were safe ; and the old marriage dates, with their subsequent desertions, were struck from the human cal-
endar. So arose, or at least was inaugurated, Family
Life, the first and the last nursery of the higher sympathies, and the home of all that was afterwards holy
in the world. One could not find a simpler instance
of the growing sovereignty of Mind over the powers
of Nature. So remote a cause as the inclination of
the earth's axis, and the consequent changes of the
seasons, determines the time of Marriage for almost
the whole animal creation, while Man, and a few other
forms of life whose environment is exceptional, are
able to refuse all such dictations. It was when Man's
mind became capable of making its own provisions
against the weather and the crops that the possibility
of Fatherhood, Motherhood, and the Family were realized.

The supporters of the hypothesis of promiscuity
have tried to show, what would almost follow from
their theory, that the children in primitive times belonged rather to the tribe. But it is not likely that
this was the case. The hypothesis of promiscuity
itself, notwithstanding its support from M'Lennan,
Morgan, Lubbock, Bastian, Post and other authorities,
has probably·received its deathblow; and the ancient-
ness of the family as well as of the institution of Marriage are both vindicated by later facts. " Everywhere," writes Westermarck, " we find the tribes or
clans composed of several families, the members of
each family being more closely connected with one
another than with the rest of the tribe. The Family,
consisting of parents, children, and often also their
next descendants, is a universal institution among ex-

isting people. And it seems extremely probable that among our early human ancestors, the Family proved, if not the Society itself, at least the nucleus of it. I do not, of course, deny that the tie which bound the children to the Mother was much more intimate and more lasting than that which bound them to the Father; but it seems to me that the only result to which a critical investigation of facts can lead us is, that in all probability there has been no stage of human development where marriage has not existed, and that the father has always been, as a rule, the protector of his Family." [1]

But the process is not yet quite completed. With the longer time together husband and wife may get to know and lean upon one another a little, but the time is still too short for deep affection, and there remain one or two serious obstacles to remove. Indeed, unless some further steps are taken, this first achievement must end in failure. As a matter of fact, it has often ended in failure, and there have been and still are tribes and nations where love between husband and wife is non-existent. Among the Hovas, we are assured by authorities, the idea of love between husband and wife is "hardly thought of"; that at Winnebah "not even the appearance of affection" exists between them; that among the Beni-Amer it is " considered even disgraceful for a wife to show any affection for her husband"; that the Chittagong Hill tribes have "no idea of tenderness nor of chivalrous devotion"; and that the Eskimo treat their wives "with great coldness and neglect." The savage **cruelty** with which wives are treated by the Aus-

[1] *Op. cit.*, pp. 42-50.

tralian aborigines is indicated even in their weapons. The very names " Servant, Slave," by which the Brahman address their wives, and the wife's reply, " Master, Lord," symbolize the gulf between the two. There are exceptions, it is true, and often touching exceptions. Travellers cite instances of constancy among savage peoples which reach the region of romance. Probably there never was a time, indeed, nor a race, when some measure of sympathy did not stir between husband and wife. But when we consider all the facts, it is impossible to doubt that in the region of all the higher affections the savage wife and the savage husband were all but strangers to each other.

What then was wanting for the perfecting of the domestic tie, and how did Evolution secure it? In the animal creation, we have already witnessed the methods which Nature took to get more care out of little care, to make a short-lived sympathy grow into a great sympathy. Her method was first, concentration; and second, extension of time. By giving a Mother one or two young to care for instead of a hundred, she made care practicable, and by lengthening the period of infancy from hours to years she made it inevitable. And these are again her methods in perfecting love between man and wife. By abolishing the pairing season she lengthened the time for love to grow in; the next step is to perfect the object on which it shall focus. For there was again the same sort of barrier to a full-blown love which we saw before in the animal kingdom. An animal mother could not truly love in the early days because she had a hundred or a thousand young. Man could

not love in the early days because he had a dozen
wives. This love was too diluted to come to anything. What Evolution next worked at was to get
a quintessence. Polygamy, in other words the scattered love of many, must, from this time forward, be
changed into monogamy—the absorbing love of one.
And this transposition was gradually introduced. A
few polygamous people, a very few at first, become
monogamous. The new system worked better, it
spread, and was finally adopted by those higher
nations which it also helped to create. It is an
instance, nevertheless, of the slowness with which
radical changes succeed in leaving great masses of
mankind, that the older system, with the ban of
Evolution upon it, still survives in Modern Europe.
Yet there are signs, even among the uncivilized, that
polygamy is passing away. Among some almost
savage tribes it is unknown; among others prohibited.
Even in a polygamous community it is usually only a
minority who have more wives than one. And where,
the plural system is in full force, the tendency—the
Evolutionist would say the transition—to monogamy
is plainly marked, for among the many wives possessed by any individual, there is generally one who
is first favorite and ranks as helpmeet or wife. The
stress just laid upon the ethical gains of the monogamous state as contrasted with the polygamous, of
course only emphasizes one side of the question, and
by the pure naturalist might be ruled out of court.
Were the physiologist to go over the same ground he
could give a parallel account of the development, and
show that on the merely physiological plane the transition to monogamy and the rise of the Family was

a likely if not an inevitable result. It is at least certain that during those later stages of social Evolution in which Monogamy has prevailed, the change has been in the best physical interests alike of the parents, the offspring, and of society.

This barrier removed, Evolution had still much to do to the other—the brevity of the time during which husband and wife remained together. What short work Nature had already made of this obstacle—by abolishing the pairing season—we have just seen. But that requires supplementing. It is not enough to give time for mutual knowledge and affection *after* marriage. Nature must deepen the result by extending it to the time *before* marriage. In primitive times there was no such thing as courtship. Men secured their wives in three ways, and in uncivilized nations so find them still. Among barbarous nations marriage is not a case of love, but of capture ; among the semi-barbarous it is a case of barter ; and among the imperfectly civilized—among whom we must often include ourselves—a matter of convention. The second of these, the purchase system—a slightly evolved form of marriage by capture—is probably one through which all human Marriage has passed ; and relics of it still exist in the *dos* and other symbols among nations with whom the custom of buying a bride has long since passed away. By degrading the object of barter to the level of a chattel, this system is a barrier to high affection. But in most cases this is heightened by the impossibility of that preliminary courtship which leads to mutual knowledge and intelligent love. The bride and bridegroom, in the extremer cases, meet as total strangers ; and though affection may bud in after years, the mingling of unknown temperaments, together with the destruction of reverence for woman by treating her as an article of barter, make the chances small of it blossoming into a flower.

Courtship, with its vivid perceptions and quickened emotions, is a great opportunity for Evolution; and to institute and lengthen reasonably a period so rich in impression is one of its latest and highest efforts. To give love time, indeed, has been all along, and through a great variety of arrangements, the chief means of establishing it on the earth. Unfortunately, the lesson of Nature here is being all too slowly learned even among nations with its open book before them. In some of the greatest of civilized countries, real mutual knowledge between the youth of the sexes is unattainable; marriages are made only by a higher kind of purchase, and the supreme step in life is taken in the dark. Whatever safeguards this method provides it cannot be final, nor can those nations rise to any exalted social height or moral greatness till some change occurs. It has been given especially to one nation to lead the world in its assault upon this mistaken law, and to demonstrate to mankind that in the unconstrained and artless relations of youth lie higher safeguards than the polite conventions of society can afford. The people of America have proved that the blending of the sweet currents of different family-lives in social intercourse, in recreation, and—most original of all—in education, can take place freely and joyously without any sacrifice of man's reverence for woman, or woman's reverence for herself; and, springing out of these naturally mingled lives, there must more and more come those sacred and happy homes which are

the surest guarantees for the moral progress of a
nation. So long as the first concern of a country is
for its homes, it matters little what it seeks second or
third. Long before Evolution showed its scientific
interest in this first social aggregate, and proclaimed
it the strategic point in moral progress, poetry, philosophy, and history assigned the same great place to
Family-life. The one point, indeed, where all students
of the past agree, where all prophets of the future
meet, where all the sciences from biology to ethics are
enthusiastically at one, is in their faith in the im-
perishable potentialties of this yet most simple institution.

With all these barriers removed it might now be
supposed that the process was at last complete.
But one of the surprises of Evolution here awaits
us. All the arrangements are finished to fan the flame
of love, yet out of none of them was love itself begotten. The idea that the existence of sex accounts
for the existence of love is untrue. Marriage among
early races, as we have seen, has nothing to do with
love. Among savage peoples the phenomenon everywhere confronts us of wedded life without a grain of
love. Love then is no necessary ingredient of the sex
relation ; it is not an outgrowth of passion. Love is
love, and has always been love, and has never been
anything lower. Whence, then, came it ? If neither
the Husband nor the Wife bestowed this gift upon the
world, Who did ? It was A Little Child. Till this
appeared, Man's affection was non-existent ; Woman's
was frozen. The Man did not love the Woman ; the
Woman did not love the Man. But one day from its
Mother's very heart, from a shrine which her husband
20

never visited nor knew was there, which she herself
dared scarce acknowledge, a Child drew forth the first
fresh bud of a Love which was not passion, a Love
which was not selfish, a Love which was an incense
from its Maker, and whose fragrance from that hour
went forth to sanctify the world. Later, long later,
through the same tiny and unconscious intermediary,
the father's soul was touched. And one day, in the
love of a little child, Father and Mother met.

That this is the true lineage of love, that it has
descended not from Husbands and Wives but through
children, is proved by the simplest study of savage
life.   Love for children is always a prior and a
stronger thing than love between Father and Mother.
The indifference of the Husband to his Wife—though
often greatly exaggerated by anthropology—is all too
manifest, and throughout whole regions the Wife does
not love but only fears her Husband.   For the
children on the other hand both parents have almost
always a regard.   The universality of a Mother's Love
is one of the revelations of travel.   Even among
cannibals, where the shocking treatment of Wives by
their Husbands is in daily evidence, a case of cruelty
to children from the Mother's side—apart from in-
fanticide which has a rationale of its own—is rarely
heard of.   The status of children if not ideal forms a
most striking contrast to the general moral and social
level : and one cannot but decide that they have been
unconsciously the true moral teachers of the world.
Had the institution of the Family depended on Sex
and not on affection it would probably never have
endured for any time.   Love is eternal ; Sex, tran-
sient.   Its unbridled expression in individual natures,

and its recklessness when thwarted, have given rise
to exaggerated ideas of its power. In all uncontrolled
forms, however, it becomes so immediate a menace to
social order, that if it does not die out in self-destruc-
tion it is checked by the community and forced into
lawful channels. The only thing that could bear the
heavy burden of social order and adapt itself to every
change and fresh demand was the indestructibly solid
yet elastic, strength of love. The care and culture
of love therefore became thenceforth the first great
charge of Evolution, and every obstruction to its path
began to be swept away. Whatever facilities could
further its career were gradually adopted, and changes
which soon began to pass over the face of all human
societies seemed but parts of one great conspiracy to
hasten its final reign.

For a prolonged and protective Fatherhood, once
introduced into the world, was immediately taken
charge of by Natural Selection. The children who
had fathers to fight for them grew up ; those which
had not, were killed or starved. The lengthening of
the period during which Father and Mother kept together meant double protection for the little ones ;
and the more they kept together for the first few
days or weeks, and the more the Father helped to
defend mother and child, the more chance for all three
in the end. The picture which Koppenfells draws of
the female Gorilla and her young ensconced in a nest
upon the fork of a tree, while Gorilla *père* sat all
night at the foot with his back against the trunk
to protect them from the leopards, is a fair object-
lesson in the first or protective stage of the Father's
Evolution. When Man passed, however, as he prob-

ably did, from the frugivorous to the carnivorous state, the Father had the additional responsibility of keeping his family in food. It would be impossible for a Mother to hunt for game and attend to her young; and for a long time the young themselves were useless in the chase, and must be entirely dependent on their parents' bounty. But this means promotion to the Father. He is not only protector but food-provider. It is impossible to believe that in process of time the discharge of this office did not bring some faint satisfactions to himself, that the mere sight of his offspring fed instead of famished did not give him a certain pleasure. And though the pleasure at first may have been no more than the absence of the annoyance they caused by the clamorousness of their want, it became a stimulus to exertion, and led in the end to rudimentary forms of sympathy and self-denial.

Once established in the world as a winning force, love could only yield to a greater force than itself and greater force there is none. In the hands of Natural Selection, therefore, it ran its course. Whatever physiological adjustments continued to go on beneath the surface, ethical factors now determined extinction or survival. Bad parents mean starved children, and starved children will be replaced in the Struggle for Life by full-fed children, and ere a few generations parents without love will exist no more. The child, on the other hand, which has drunk most deeply of its Father's or its Mother's love lives to hand on that which has spared it to a succeeding race. How much of affection is handed on, or how little, matters not, for Heredity works with the finest microscope, and sees, and seizes, the invisible. In a second child, reared by parents one degree more loving than the last, this ultimate particle of love will grow a little more, and each succeeding Family in this royal line will be richer in the elements which make for progress than the last.

When we reach the human Family, we find that this simple combination was already strong enough to become the nucleus of the social and national life of the world. For the moment the new forces of Sympathy, Brotherhood, Self-denial, or Love, began to work among the isolated units which made up primitive Man, the whole composition and character of the aggregate began to change. Sooner or later in the recurring necessities of savage existence there came an opportunity for the members of the first combination, the little group of Father, Mother, and Sons, to act together. However unworthily this primitive group merited the name of Family, there was here what at that time was of final importance—the elements of physical strength. He who formerly stood alone in the Struggle for Life now found himself backed on occasion by an inner circle. Those who outside this circle ventured to oppose or offend an individual within it had the Family to reckon with. Ends were gained by the new alliance which were unattainable single-handed by any individual member of the tribe, and whether enlisted to evade disaster or secure a prey, to resist an injustice or avenge a wrong, the odds henceforth and always were in favor of the combination. When it is remembered how, owing to the comparative equality of the competitors in the conflict of savage existence, even an infinitesimal

advantage on one side or the other determines health
or starvation, survival or extinction, the importance
of the first feeble effort at federation must be recognized. Shoulder to shoulder has been the watchword
all through history of national development. Almost
from the very first, indeed, the Family and not the individual must have been the unit of Tribal life ; and as
Families grew more and more definite, they became
the recognized piers of the social structure and gave
a first stability to the race of men.

But great as are the physical advantages of the
Family, the ethical uses, even in the early days of its
existence, place this institution at the head of all the
creations of Evolution. For the Family is not only
its greatest creation, but its greatest instrument for
further creation. The ethical changes begin almost
the moment it is formed. One immediate effect, for
instance, of the formation of Family groups was to
take off from any single individual the perpetual
strain of the Struggle for Life. The Family as a
whole must sometimes fight, but the responsibility
and the duty are now distributed, and those who were
once solely pre-occupied with the personal struggle
will have respites, during which other things will
occupy their minds. Attention thus called off from
environing enemies, the members of the Family will,
as it were, discover one another. New relations
among them will spring up, new adjustments to one
another's presence and to one another's needs, and
hitherto unknown elements of character will be gradually called to the surface. That unselfishness, in
some rude form, should now grow up is a necessity of
living together. A man cannot be a member of a

Family and remain an utter egoist. His interests are perforce divided, and though the Family group is a small surface for unselfishness to spread to and to practise on, no greater feat could as yet be attempted, and Evolution never runs risks of too rapid development or over-strain. With the incorporation of the Family into a Clan or Tribe the area will presently be extended, and the necessity of controlling self-interest more thoroughly, or merging it in a wider interest, become more obligatory. But to prepare the altruistic sentiment for so great an abnegation, the simpler discipline of the Family was required. How firmly Families in time became welded together in mutual interest and support, and how much crude Altruism this implies, is evident from the place of Family feuds and the power of great Families and Houses both in ancient and modern history. A striking instance is the Vendetta. To avenge a Family insult in countries where this prevails was a sacred duty to all the relatives, and even the last surviving member willingly gave up his life to vindicate its honor. So strong indeed sometimes has grown the power of individual Families that the more desirable spread of Altruism to the Nation was threatened, and wider interests so much forgotten that the Family became the enemy of the State. Nothing could more forcibly show the tremendous power of self-development contained within the Family circle, and the solidity and strength to which it can grow, than that, time after time in history, it has had to be crushed and broken up by all the forces of the State.

Among other elements in human nature fostered in

the Family is one of exceptional interest. The attempt has been made to show that from the inevitable
relations of early Family life, the sense of Duty first
dawned upon the world. The theme is too great, too
intricate, and too dangerous to open under the limitations of the present inquiry, for these deny us the
appeal to Society, to Religion, and even to the Conscience of the higher Man. But it is due to the
Father, whose Evolution we are tracing, that the
share he is supposed by some authorities to take in it,
should be at least named.

That morality in general has something to do with
the relations of people to one another is evident, as
every one knows, from the mere derivation of the
word. *Mores*, morals, are in the first instance *customs*, the customs or ways which people have when
they are together. Now, the Family is the first occasion of importance where we get people together.
And as there are not only a number of people in a
Family, but different kinds of people, there will be a
variety in the relations subsisting between them, in
the customs which stereotype the most frequently repeated actions necessitated by these relations, and in
the moods and attitudes of mind accompanying them.
Leaving out of sight differences of kind among broth-
ers and sisters, consider the probably more divergent
and certainly more dominant influences of Father and
Mother. What the relation of child to Mother has
crystallized into we have sufficiently marked—it is a
relation of direct dependence, and its product is Love.
But the Father is a wholly different influence. What
attitude does the Child take up in this austerer presence, and what ways of acting, what customs, *mores,*

morals, are engrained in the child's mind through it?
The acknowledged position of the Father in most early
tribes is head of the Family. To the children, and
generally even to the Mother, he represents Authority. He is the children's chief. Bachoven has famil-
iarized us with the idea of a Matriarchate, or Maternal
Family; but although exceptional tribes have given
supremacy to the Mother, the rule is for the Father to
be supreme. As head of the Family, therefore, it was
his business to make the Family laws. No doubt the
Mother also made laws; but the Father, as the more
terrible person, exacted obedience with greater severity, and his laws acquired more force. To do what
was pleasing in his eyes was a necessity with the
children, and his favor or his frown became standards
of what was " good " and what was " bad." Low as
this standard was—the fear or favor of a savage
Father—it was a beginning of right *mores*, good conduct, proper manners. Plant in the mind, or evoke
from it, the idea of acting in a given way with reference to some half-dozen daily trifles when done in the
presence of one authoritative individual, and Evolution has already found something to work on. The
children have got six, if not ten commandments. Extend the half-dozen things done rightly to a whole
dozen, and then to a score, and then to a hundred;
and let it become habitual to do these things rightly.
When the right doing of these things commends the
doer to one person, he will next be apt to commend
himself by similar conduct to other persons, if their
standard happens to be the same. Whether good behavior purchases favor or simply succeeds in evading
penalties is at first immaterial. All that is required,

under whatever sanctions, is that some standard of good or bad shall arise. No abstract sense of duty, of course, here exists ; no perfect law ; it is a purely personal and local code ; but the word duty has at least received a first imperfect meaning; and the Father, in some rough way, forms an external conscience to those beneath him.

Such is the tentative theory of the advocates of Evolutional Ethics. It may or may not be a possible account of the rise of a sense of obligation, but it is certain that it does not account for the whole of it. Why, also, that particular thing should be elicited under the circumstances described is an unanswered question. In attempting to trace its rise, no rationale appears of its origin; all proofs, in short, of its evolution take for granted its previous existence. A latent thing has become active; an invisible thing has become apparent. In one sense a relation has been created, in another sense a quality in that relation has been revealed. A new experiment upon human nature has been tried; a new discovery of its properties has been the result.

That these moral elements, on the other hand, must have a beginning somewhere in space and time is certain enough. Not less necessary to the world than the Mother's gift of Love is the twin offering of the Father—Righteousness. And if, almost before the soul is born, the shadowy outline of a moral order should begin to loom out in history, the later phases and the later sanctions lose nothing of their quality, are all the more wonderful and all the more divine, because met by moral adumbrations in the distant past. If the later children had their ten commandments given them in one way, they cannot grudge that the world's earlier children should have been given their two or three commandments in another way—another way which, nevertheless, did we know all, might turn out to be but another phase of the same way. But it is impossible even to approach the Evolution of Morality until we have carried Man some stages further up his Ascent. It is only when he reaches the social stage, when he becomes aggregated into clans, tribes, and nations, that this problem opens. For the present we must content ourselves with having witnessed his arrival in the Human Family—the starting-point and threshold of the true moral life.

For a long time, it is true, the Family circle, as a circle, was incomplete. Machinery must itself evolve before its products evolve. Scarcely defined at all, broken as soon as formed, the earlier circles allowed their strongest forces to escape almost at the moment they generated. But the walls grew higher and higher with the advance of history. The leakage became less and less. With the Christian era the machinery was complete; the circle finally closed in, and became a secluded shrine where the culture of everything holy and beautiful was carried on. The path by which this ideal consummation was reached was not, as we have seen, a straight path; nor has the integrity of the institution been always preserved through the later centuries. The difficulty of realizing the ideal may be judged of by the fewness of the nations now living who have reached it, and by the multitude of peoples and tribes who have vanished from the earth without attaining. From the failure

to fulfil some one or other of the required conditions
people after people and nation after nation have come
together only to disperse, and leave no legacy behind
except the lesson—as yet in few cases understood—of
why they failed.

Yet whether the road be straight or devious is of
little moment. The one significant thing is that it
rises. We have reached a stage in Evolution at
which physiological gains are guarded and accentuated, if not in an ethical interest, at least by ethical factors becoming utilized by natural selection.
Henceforth affection becomes a power in the world;
and whatever physiological adjustments continue to
go on beneath the surface, the most attached Families
will have a better chance of surviving and of trans-
mitting their moral characteristics to succeeding generations. The completion of the arch of Family Life
forms one of the great, if not the greatest of the landmarks of history. If the crowning work of Organic
Evolution is the Mammalia; the consummation of
the Mammalia is the Family. Physically, psychically, ethically, the Family is the masterpiece of
Evolution. The creation of Evolution, it was destined
to become the most active instrument and ally which
Evolution has ever had. For what is its evolutionary
significance? It is the generator and the repository of
the forces which alone can carry out the social and
moral progress of the world. There they rally when
they become enfeebled, there their excesses are coun-
terbalanced, and thence they radiate out, refined and
reinforced, to do their holy work.

Looking at the mere dynamics of the question, the
Family contains all the machinery, and nearly all the

power, for the moral education of mankind. Feebly, but adequately, in the early chapters of Man's history it fulfilled its function of nursing Love, the Mother of all morality; and Righteousness, the Father of all morality, so preparing a parentage for all the beautiful spiritual children which in later years should spring from them. If life henceforth is to go on at all, it must be a better life, a more loving life, a more abundant life; and this premium upon Love means— if it means anything—that Evolution is taking henceforth an ethical direction. It is no more possible to interpret Nature physically from this point than to interpret a "Holy Family" of Raphael's in terms of the material structure of canvas or the qualities of pigments. Canvas may be coarse or fine, pigments may be vegetable or mineral; but whether the colors be crushed out of madder or ground out of arsenic or lead is of no importance now. Once these things were important; by infinitely slow processes Nature formed them; by clever arts the colorman prepared them. But the "Holy Family" did not lie potentially in the madder-bud, nor in the earth with the lead and arsenic, nor in the laboratory with the colorman. He who claims Nature for Matter and Physical force makes the same assumption that these would do if they claimed the painting. In a far truer sense than Raphael produced his "Holy Family" Nature has produced a Holy Family. Not for centuries but for millenniums the Family has survived. Time has not tarnished it; no later art has improved upon it; nor genius discovered anything more lovely; nor religion anything more divine. From the bee's cell and the butterfly's wing men draw what they call the Argu-

ment from Design ; but it is in the kingdoms which
come without observation, in these great immaterial
orderings which Science is but beginning to perceive,
hat the purposes of Creation are revealed.

CHAPTER X.

INVOLUTION.

MANY years ago, in the clay which in every part of the world is found underlying beds of coal, a peculiar fossil was discovered and named by science Stigmaria. It occurred in great abundance and in many countries, and from the strange way in which it ramified through the clay it was supposed to be some extinct variety of a gigantic water-weed. In the coal itself another fossil was discovered, almost as abundant but far more beautiful, and from the exquisite carving which ornamented its fluted stem it received the name of Sigillaria. One day a Canadian geologist, studying Sigillaria in the field, made a new discovery. Finding the trunk of a Sigillaria standing erect in a bed of coal, he traced the column downwards to the clay beneath. To his surprise he found it ended in Stigmaria. This branching fossil in the clay was no longer a water-weed. It was the root of which Sigillaria was the stem, and the clay was the soil in which the great coal-plant grew.

Through many chapters, often in the dark, everywhere hampered by the clay, we have been working among roots. Of what are they the roots ? To what order do they belong? By what process have they grown? What connection have they with the realm above, or the realm beneath? Is it a Stigmaria or a Sigillaria world?

Till yesterday Science did not recognize them even as roots. They were classified apart. They led to nothing. No organic connection was known between lower Nature and that wholly separate and all but antagonistic realm, the higher world of Man. Atoms, cells, plants, animals were the material products of a separate creation, the clay from which Man took his clay-body, and no more. The higher world, also, was a system by itself. It rose out of nothing; it rested upon nothing. Clay, where the roots lay, was the product of inorganic forces; Coal, which enshrined the tree, was a creation of the sunlight. What fellow-ship had light with darkness? What possible connection could exist between that beautiful organism which stood erect in the living, and that which lay prone in the dead? Yet, by a process doubly verified, the organic connection between these two has now been traced. Working upwards through the clay the biologist finds what he took to be an organism of the clay leaving his domain and passing into a world above—a world which he had scarcely noticed before, and into which, with such instruments as he employs, he cannot follow it. Working downward through the higher world, the psychologist, the moralist, the sociologist, behold the even more wonderful spectacle of the things they had counted a peculiar possession of the upper kingdom, burying themselves in ever attenuating forms in the clay beneath. What is to be made of this discovery? Once more, Is it a Stigmaria or a Sigillaria world? Is the biologist to give up his clay or the moralist his higher kingdom? Are Mind, Morals, Men, to be interpreted in terms of roots, or are atoms and cells to be judged by the flowers and fruits of the tree?

The first fruit of the discovery must be that each shall explore with new respect the other's world, and, instead of delighting to accentuate their contrasts, strive to magnify their infinite harmonies. Old as is the world's vision of a cosmos, and universal as has been its dream of the unity of Nature, neither has ever stood before the imagination complete. Poetry felt, but never knew, that the universe was one; Biology perceived the profound chemical balance between the inorganic and organic kingdoms, and no more; Physics, discovering the correlation of forces, constructed a cosmos of its own; Astronomy, through the law of gravitation, linked us, but mechanically, with the stars. But it was reserved for Evolution to make the final revelation of the unity of the world, to comprehend everything under one generalization, to explain everything by one great end. Its omnipresent eye saw every phenomenon and every law. It gathered all that is and has been into one last whole —a whole whose very perfection consists in the all but infinite distinctions of the things which it unites.

What is often dreaded in Evolution—the danger of obliterating distinctions that are vital—is a groundless fear. Stigmaria can never be anything more than root, and Sigillaria can never be anything less than stem. To show their connection is not to transpose their properties. The wider the distinctions seen among their properties the profounder is the Thought
21

which unites them, the more rich and rational the Cosmos which comprehends them. For "the unity which we see in Nature is that kind of Unity which the Mind recognizes as the result of operations similar to its own—not a unity which consists in mere sameness of material, or in mere identity of composition, or in mere uniformity of structure; but a unity which consists in the subordination of all these to similar aims, not to similar principles of action—that is to say, in like methods of yoking a few elementary forces to the discharge of special functions, and to the production, by adjustment, of one harmonious whole." [1]

Yet did Sigillaria grow out of Stigmaria? Did Mind, Morals, Men, evolve out of Matter? Surely if one is the tree and the other the root of that tree, and if Evolution means the passage of the one into the other, there is no escape from this conclusion—no escape therefore from the crassest materialism? If this is really the situation, the lower must then include the higher, and Evolution, after all, be a process of the clay? This is a frequent, a natural, and a wholly unreflecting inference from a very common way of stating the Evolution theory. It arises from a total misconception of what a root is. Because a thing is seen to have roots, it is assumed that it has grown out of these roots, and must therefore belong to the root-order. But neither of these things is true in Nature. Are the stem, branch, leaf, flower, fruit of a tree roots? Do they belong to the root-order? They do not. Their whole morphology is different; their whole physiology is different; their reactions upon the world around are different. But it must be allowed that

[1] Duke of Argyll, *The Unity of Nature*, p. 44.

they are at least contained in the root? No single one
of them is contained in the root. If not in the root,
then in the clay? Neither are they contained in the
clay. But they grow out of clay, are they not made
out of clay? They do not grow out of clay, and they
are not made out of clay. It is astounding sometimes
how little those who venture to criticise biological
processes seem to know of its simplest facts. Fill a
flower-pot with clay, and plant in it a seedling. At the
end of four years it has become a small tree ; it is six
feet high ; it weighs ten pounds. But the clay in the
pot is still there ? A moiety of it has gone, but it is
not appreciably diminished ; it has not, except the
moiety, passed into the tree ; the tree does not live on
clay nor on any force contained in the clay. It cannot
have grown out of the seedling, for the seedling contained
but a grain for every pound contained in the tree. It
cannot have grown from the root, because the root is
there now, has lost nothing to the tree, has itself gained
from the tree, and at first was no more there than
the tree.

Sigillaria, then, as representing the ethical order,
did not grow out of Stigmaria as representing the
organic or the material order. Trees not only do not
evolve out of their roots, but whole classes in the
plant world—the sea-weeds for instance—have no roots
at all. If any possible relation exists it is exactly
the opposite one—it is the root which evolves from the
tree. Trees send down roots in a far truer sense than
roots send up trees. Yet neither is the whole truth.
The true function of the root is to give stability to the
tree, and to afford a medium for conveying into it
inorganic matter from without. And this brings us

face to face with the real relation. Tree and root—the seed apart—find their explanation not in one another nor in something in themselves, but mainly in something outside themselves. The secret of Evolution lies, in short, with *the Environment*. In the Environment, in that in which things live and move and have their being, is found the secret of their being, and especially of their becoming. And what is that in which things live and move and have their being ? It is Nature, the world, the cosmos—and something more, some One more, an Infinite Intelligence and an Eternal Will. Everything that lives, lives in virtue of its correspondences with this Environment. Evolution is not to unfold from within; it is to infold from without. Growth is no mere extension from a root but a taking possession of, or a being possessed by, an ever widening Environment, a continuous process of assimilation of the seen or Unseen, a ceaseless re-distribution of energies flowing into the evolving organism from the Universe around it. The supreme factor in all development is Environment. Half the confusions which reign round the process of Evolution, and half the objections to it, arise from considering the evolving object as a self-sufficient whole. Produce an organism, plant, animal, man, society, which will evolve *in vacuo* and the right is yours to say that the tree lies in the root, the flower in the bud, the man in the embryo, the social organism in the family of an anthropoid ape. If an organism is to be judged in terms of the immediate Environment of its roots, the tree is a clay tree; but if it is to be judged by stem, leaves, fruit, it is not a clay tree. If the moral or social organism is to be judged in terms of the Environment of its roots, the moral and social organism is a material organism; but if it is to be judged in terms of the higher influences which enter into the making of its stem, leaves, fruit, it is not a material organism. Everything that lives, and every part of everything that lives, enters into relation with different parts of the Environment and with different things in the Environment; and at every step of its Ascent it compasses new ranges of the Environment, and is acted upon, and acts, in different ways from those in which it was acted upon, or acted, at the previous stage.

For what is most of all essential to remember is that not only is Environment the prime factor in development, but that the Environment itself rises with every evolution of any form of life. To regard the Environment as a fixed quantity and a fixed quality is, next to ignoring the altruistic factor, the cardinal error of evolutional philosophy. With every step a climber rises up a mountain side his Environment must change. At a thousand feet the air is lighter and purer than at a hundred, and as the effect varies with the cause, all the reactions of the air upon his body are altered at the higher level. His pulse quickens; his spirit grows more buoyant; the energies of the upper world flow in upon him. All the other phenomena change—the plants are Alpine, the animals are a hardier race, the temperature falls, the very world he left behind wears a different look. At three thousand feet the causes, the effects, and the phenomena change again. The horizon is wider, the light intenser, the air colder, the top nearer; the nether world recedes from view. At six thousand feet, if we may accentuate the illustration till it contains more of the emphasis of the reality, he enters the region of snow. Here is a change brought about by a small and perfectly natural rise which yet amounts to a revolution. Another thousand feet and there is another revolution—he is ushered into the domain of mist. Still another thousand, and the climax of change has come. He stands at the top, and, behold, the Sun. None of the things he has encountered in his progress toward the top are new things. They are the normal phenomena of altitude—the scenes, the energies, the correspondences, natural to the higher slopes. He did not create any of these things as he rose; they were not created as he rose; they did not lie potentially in the plains or in the mountain foot. What has happened is simply that in rising he has encountered them—some for the first time, which are therefore wholly new to him; others which, though known before, now flow into his being in such fuller measure, or enter into such fresh relations among themselves, or with the changed being which at every step he has become, as to be also practically new.

Man, in his long pilgrimage upwards from the clay, passes through regions of ever-varying character. Each breath drawn and utilized to make one upward step brings him into relation with a fractionally higher air, a fractionally different world. The new energies he there receives are utilized, and in virtue of them he rises to a third, and from a third to a fourth. As in the animal kingdom the senses open one by one —the eye progressing from the mere discernment of light and darkness to the blurred image of things near, and then to clearer vision of the more remote;

the ear passing from the tremulous sense of vibration to distinguish with ever-increasing delicacy the sounds of far-off things—so in the higher world the moral and spiritual senses rise and quicken till they compass qualities unknown before and impossible to the limited faculties of the earlier life. So Man, not by any innate tendency to progress in himself, nor by the energies inherent in the protoplasmic cell from which he first set out, but by a continuous feeding and reinforcing of the process from without, attains the higher altitudes, and from the sense-world at the mountain foot ascends with ennobled and ennobling faculties until he greets the Sun.

What is the Environment of the Social tree ? It is all the things, and all the persons, and all the influences, and all the forces with which, at each successive stage of progress, it enters into correspondence. And this Environment inevitably expands as the Social tree expands and extends its correspondences. At the savage stage Man compasses one set of relations, at the rude social stage another, at the civilized stage a third, and each has its own reactions. The social, the moral, and the religious forces beat upon all social beings in the order in which the capacities for them unfold, and according to the measure in which the capacities themselves are fitted to contain them. And from what ultimate source do they come ? There is only one source of everything in the world. They come from the same source as the Carbonic Acid Gas, the Oxygen, the Nitrogen, and the Vapor of Water, which from the outer world enter into the growing plant. These also visit the plant in the order in which the capacities for them unfold, and according to the measure in which these capacities can contain them.

The fact that the higher principles come from the same Environment as those of the plant, nevertheless does not imply that they are the same as those which enter into the plant. In the plant they are physical, in Man spiritual. If anything is to be implied it is not that the spiritual energies are physical, but that the physical energies are spiritual. To call the things in the physical world "material" takes us no nearer the natural, no further away from the spiritual. The roots of a tree may rise from what we call a physical world; the leaves may be bathed by physical atoms; even the energy of the tree may be solar energy, but the tree is *itself*. The tree is a Thought, a unity, a rational purposeful whole; the "matter" is but the medium of their expression. Call it all—matter, energy, tree—a physical production, and have we yet touched its ultimate reality? Are we even quite sure that what we call a physical world is, after all, a physical world? The preponderating view of science at present is that it is not. The very term "material world," we are told, is a misnomer; that the world is a spiritual world, merely employing "matter" for its manifestations.

But surely there is still a fallacy. Are not these so-called social forces, the effect of Society and not its cause? Has not Society to generate them before they regenerate Society? True, but to generate is not to create. Society is machinery, a medium for the transmission of energy, but no more a medium for its creation than a steam engine is for the creation of its energy. Whence then the social energies? The answer is as before. Whence the physical energies? And Science has only one answer to that. "Consider the position into which Science has brought us. We are led by scientific logic to an unseen, and by scientific analogy to the spirituality of this unseen. In fine, our conclusion is, that the visible universe has been developed by an intelligence resident in the Unseen."[1] There is only one theory of the method of Creation in the field, and that is Evolution; but there is only one theory of origins in the field, and that is Creation. Instead of abolishing a creative Hand, Evolution demands it. Instead of being opposed to Creation, all theories of Evolution begin by assuming it. If Science does not formally posit it, it never posits anything less. "The doctrine of Evolution," writes Mr. Huxley, "is neither theistic nor anti-theistic. It has no more to do with theism than the first book of Euclid has. It does not even come in contact with theism considered as a scientific doctrine." But when it touches the question of origins, it is either theistic or silent. "Behind the co-operating forces of Nature," says Weismann, "which aim at a purpose, we must admit a cause, . . . inconceivable in its nature, of which we can only say one thing with certainty, that it must be theological."

The fallacy of the merely quantitative theory of Evolution is apparent. To interpret any organism in terms of the organism solely is to omit reference to the main instrument of its Evolution, and therefore to leave the process, scientifically and philosophically,

[1] Balfour Stewart and Tait, *The Unseen Universe*, 6th edition, p. 221.

unexplained. It is as if one were to construct a theory of the career of a millionaire in terms of the pocket-money allowed him when a schoolboy. Disregard the fact that more pocket-money was allowed the schoolboy as he passed from the first form to the sixth; that his allowance was increased as he came of age; that now, being a man, not a boy, he was capable of more wisely spending it; that being wise he put his money to paying uses; and that interest and capital were invested and re-invested as years went on—disregard all this and you cannot account for the rise of the millionaire. As well construct the millionaire from the potential gold contained in his first sixpence—a sixpence which never left his pocket—as construct a theory of the Evolution of Man from the protoplasmic cell apart from its Environment. It is only when interpreted, not in terms of himself, but in terms of Environment, and of an Environment increasingly appropriated, quantitatively and qualitatively, with each fresh stage of the advance, that a consistent theory is possible, or that the true nature of Evolution can appear.

A child does not grow out of a child by spontaneous unfoldings. The process is fed from without. The body assimilates food, the mind assimilates books, the moral nature draws upon affection, the religious faculties nourish the higher being from Ideals. Time brings not only more things, but new things; the higher nature inaugurates possession of, or by, the higher order. "It lies in the very nature of the case that the earliest form of that which lives and develops is the least adequate to its nature, and therefore that from which we can get the least distinct clue to the inner principle of that nature. Hence to trace a living being back to its beginning, and to explain what follows by such beginning, would be simply to omit almost all that characterizes it, and then to suppose that in what remains we have the secret of its existence. That is not really to explain it, but to explain it away; for on this method, we necessarily reduce the features that distinguish it to a *minimum*, and, when we have done so, the remainder may well seem to be itself reducible to something in which the principle in question does not manifest itself at all. If we carry the animal back to protoplasm, it may readily seem possible to explain it as a chemical compound. And, in like manner, by the same minimizing process, we may seem to succeed in reducing consciousness and self-consciousness in its simplest form to sensation, and sensation in its simplest form to something not essentially different from the nutritive life of plants. The fallacy of the *sorites* may thus be used to conceal all *qualitative* changes under the guise of quantitative addition or diminution, and to bridge over all difference by the idea of gradual transition. For, as the old school of etymologists showed, if we are at liberty to interpose as many connecting links as we please, it becomes easy to imagine that things the most heterogeneous should spring out of each other. While, however, the hypothesis of gradual change—change proceeding by infinitesimal stages which melt into each other so that the eye cannot detect where one begins and the other ends—makes such a transition easier for *imagi-*

nation, it does nothing to diminish the difficulty or the wonder of it for *thought.*" [1]

The value of philosophical criticism to science has seldom appeared to more advantage than in these words of the Master of Balliol. The following passage from Martineau may be fitly placed beside them :—
" In not a few of the progressionists the weak illusion is unmistakable, that, with time enough, you may get everything out of next to nothing. Grant us, they seem to say, any tiniest granule of power, so close upon zero that it is not worth begrudging—allow it some trifling tendency to infinitesimal movement—and we will show you how this little stock became the kosmos, without ever taking a step worth thinking of, much less constituting a case for design. The argument is a mere appeal to an incompetency in the human imagination, in virtue of which magnitudes evading conception are treated as out of existence; and an aggregate of inappreciable increments is simultaneously equated,—in its cause to *nothing*, in its effect to *the whole of things.* You manifestly want the same causality, whether concentrated in a moment or distributed through incalculable ages; only in drawing upon it a logical theft is more easily committed piecemeal than wholesale. Surely it is a mean device for a philosopher thus to crib causation by hairs-breadths, to put it out at compound interest through all time, and then disown the debt." [2]

It is not said that the view here given of the process of Evolution has been the actual process. The illustrations have been developed rather to clear up dif-

[1] Edward Caird, *The Evolution of Religion,* Vol. I., pp. 49–50.
[2] Martineau, *Essays, Philosophical and Theological,* p. 141.

ficulties than to state a theory. The time is not ripe for daring to present to our imaginations even a partial view of what that transcendent process may have been. At present we can only take our ideas of growth from the growing things around us, and in this analogy we have taken no account of the most essential fact—the seed. Nor is it asserted, far as these illustrations point in that direction, that the course of Evolution has been a continuous, uninterrupted, upward rise. On the whole it has certainly been a rise; but whether a rise without leap or break or pause, or—what is more likely—a progress in rhythms, pulses, and waves, or—what is unlikely—a cataclysmal ascent by steps abrupt and steep, may possibly never be proved.

There are reverent minds who ceaselessly scan the fields of Nature and the books of Science in search of gaps—gaps which they will fill up with God. As if God lived in gaps? What view of Nature or of Truth is theirs whose interest in Science is not in what it can explain but in what it cannot, whose quest is ignorance not knowledge, whose daily dread is that the cloud may lift, and who, as darkness melts from this field or from that, begin to tremble for the place of His abode? What needs altering in such finely jealous souls is at once their view of Nature and of God. Nature is God's writing, and can only tell the truth; God is light, and in Him is no darkness at all.

If by the accumulation of irresistible evidence we are driven—may not one say permitted—to accept Evolution as God's method in creation, it is a mistaken policy to glory in what it cannot account for. The reason why men grudge to Evolution each of its fresh

claims to show how things have been made is the
groundless fear that if we discover· how they are made
we minimize their divinity. When things are known,
that is to say, we conceive them as natural, on Man's
level; when they are unknown, we call them divine—
as if our ignorance of a thing were the stamp of its
divinity. If God is only to be left to the gaps in our
knowledge, where shall we be when these gaps are
filled up? And if they are never to be filled up, is
God only to be found in the disorders of the world?
Those who yield to the temptation to reserve a point
here and there for special divine interposition are apt
to forget that this virtually excludes God from the
rest of the process. If God appears periodically, he
disappears periodically. If he comes upon the scene at
special crises he is absent from the scene in the intervals. Whether is all-God or occasional-God the nobler theory? Positively, the idea of an immanent God,
which is the God of Evolution, is infinitely grander
than the occasional wonder-worker who is the God of
an old theology. Negatively, the older view is not
only the less worthy, but it is discredited by science.
And as to facts, the daily miracle of a flower, the
courses of the stars, the upholding and sustaining day
by day of this great palpitating world, need a living
Will as much as the creation of atoms at the first. We
know growth is the method by which things are made
in Nature, and we know no other method. We do not
know that there are not other methods; but if there
are we do not know them. Those cases which we
do not know to be growths, we do not know to be
anything else, and we may at least suspect them to
be growths. Nor are they any the less miraculous

because they appear to us as growths. A miracle is not *something quick*. The doings of these things may seem to us no miracle, nevertheless it is a miracle that they have been done.

But, after all, the miracle of Evolution is not the process, but the product. Beside the wonder of the result, the problem of the process is a mere curiosity of Science. For what is the product? It is not mountain and valley, sky and sea, flower and star, this glorious and beautiful world in which Man's body finds its home. It is not the god-like gift of Mind nor the ordered cosmos where it finds so noble an exercise for its illimitable powers. It is that which of all other things in the universe commends itself, with increasing sureness as time goes on, to the reason and to the heart of Humanity—Love. Love is the final result of Evolution. This is what stands out in Nature as the supreme creation. Evolution is not progress in matter. Matter cannot progress. It is a progress in spirit, in that which is limitless, in that which is at once most human, most rational, and most divine. Whatever controversy rages as to the factors of Evolution, whatever mystery enshrouds its steps, no doubt exists of its goal. The great landmarks we have passed, and we are not yet half-way up the Ascent, each separately and all together have declared the course of Nature to be a rational course, and its end a moral end. At the furthest limit of time, in protoplasm itself, we saw start forth the two great currents which, by their action and reaction, as Selfishness and Unselfishness, were to supply in ever accentuating clearness the conditions of the moral life. Following their movements upward through the

organic kingdom, we watched the results which each
achieved—always high, and always waxing higher;
and though what we called Evil dogged each step with
sinister and sometimes staggering malevolence, the
balance when struck, was always good upon the
whole. Then came the last great act of the organic
process, the act which finally revealed to teleology its
hitherto obscured end, the organization of the Mammalia, the Kingdom of the Mothers. So full of ethical
possibility is this single creation that one might stake
the character of Evolution upon the Mammalia alone.
On the biological side, as we have seen, the Evolution
of the Mammalia means the Evolution of Mothers; on
the sociological side, the Evolution of the Family ; and
on the moral side, the Evolution of Love. How are
we to characterize a process which ripened fruits like
these ? That the very animal kingdom had for its end
and crown a class of animals who owe their name,
their place, and their whole existence to Altruism;
that through these Mothers society has been furnished
with an institution for generating, concentrating,
purifying, and re-distributing Love in all its enduring
forms; that the perfecting of Love is thus not an incident in Nature, but everywhere the largest part of her
task, begun with the first beginnings of life, and continuously developing quantitatively and qualitatively
to the close—all this has been read into Nature by our
own imaginings, or it is the revelation of a purpose
of benevolence and a God whose name is Love. The
sceptic, **we** are sometimes reminded, has presented
crucial difficulties to the theist founded on the
doctrine of Evolution. Here is a problem which the
theist may leave with the sceptic. That that which

has emerged has the qualities it has, that even the Mammalia should have emerged, that that class should stand related to the life of Man in the way it does, that Man has lived because he loved, and that he lives to love—these, on any theory but one, are insoluble problems.

Forbidden to follow the Evolution of Love into the higher fields of history and society, we take courage to make a momentary exploration in a still lower field—a field so far beneath the plant and animal level that hitherto we have not dared to enter it. Is it conceivable that in inorganic Nature, among the very material bases of the world, there should be anything to remind us of the coming of this Tree of Life? To expect even foreshadowings of ethical characters there were an anachronism too great for expression. Yet there is something there, something which is at least worth recalling in the present connection.

The earliest condition in which Science allows us to picture this globe is that of a fiery mass of nebulous matter. At the second stage it consists of countless myriads of similar atoms, roughly outlined into a ragged cloud-ball, glowing with heat, and rotating in space with inconceivable velocity. By what means can this mass be broken up, or broken down, or made into a solid world? By two things—mutual attraction and chemical affinity. The moment when within this cloud-ball the conditions of cooling temperature are such that two atoms could combine together the cause of the Evolution of the Earth is won. For this pair of atoms are chemically "stronger" than any of the atoms immediately surrounding them. Gradually, by attraction or affinity, the primitive pair of

22

atoms—like the first pair of savages—absorb a third atom, and a fourth, and a fifth, until a "Family" of atoms is raised up which possesses properties and powers altogether new, and in virtue of which it holds within its grasp the conquest and servitude of all surrounding units. From this growing centre, attraction radiates on every side, until a larger aggregate, a family group—a Tribe—arises and starts a more powerful centre of its own. With every additional atom added, the power as well as the complexity of the combination increases. As the process goes on, after endless vicissitudes, repulsions, and readjustments, the changes become fewer and fewer, the conflict between mass and mass dies down, the elements passing through various stages of liquidity finally combine in the order of their affinities, arrange themselves in the order of their densities, and the solid earth is finished.

Now recall the names of the leading actors in this stupendous reformation. They are two in number, mutual attraction and chemical affinity. Notice these words—Attraction, Affinity. Notice that the great formative forces of physical Evolution have psychical names. It is idle to discuss whether there is or can be any identity between the thing represented in the one case and in the other. Obviously there cannot be. Yet this does not exhaust the interest of the analogy. In reality, neither here nor anywhere, have we any knowledge whatever of what is actually meant by Attraction; nor, in the one sphere or in the other, have we even the means of approximating to such knowledge. To Newton himself the very conception of one atom or one mass, attracting through empty space another atom or another mass, put his mental powers to confusion. And as to the term Affinity, the most recent Chemistry, finding it utterly unfathomable in itself, confines its research at present to the investigation of its modes of action. Science does not know indeed what forces are; it only classifies them. Here, as in every deep recess of physical Nature, we are in the presence of that which is metaphysical, that which bars the way imperiously at every turn to a materialistic interpretation of the world. Yet name and nature of force apart, what affinity even the grossest, what likeness even the most remote, could one have expected to trace between the gradual aggregation of units of matter in the condensation of a weltering star, and the slow segregation of men in the organization of societies and nations? However different the agents, is there no suggestion that they are different stages of a uniform process, different epochs of one great historical enterprise, different results of a single evolutionary law?

Read from the root, we define this age-long process by a word borrowed from the science of roots—a word from the clay—Evolution. But read from the top, Evolution is an impossible word to describe it. The word is Involution. It is not a Stigmaria world, but a Sigillaria world; a spiritual, not a material universe. Evolution is Advolution; better, it is Revelation—the phenomenal expression of the Divine, the progressive realization of the Ideal, the Ascent of Love. Evolution is a doctrine of unimaginable grandeur. That Man should discern the prelude to his destiny in the voices of the stars; that the heart of Nature should be a so human heart; that its eternal enterprise should be one with his ideals; that even in the Universe beyond, the Reason which presides should have so strange a kinship with that measure of it which he calls his own; that he, an atom in that Universe, should dare to feel himself at home within it, should stand beside Immensity, Infinity, Eternity, unaffrighted and undismayed—these things bewilder Man the more that they bewilder him so little.

But one verdict is possible as to the practical import of this great doctrine, as to its bearing upon the individual life and the future of the race. Evolution has ushered a new hope into the world. The supreme message of science to this age is that all Nature is on the side of the man who tries to rise. Evolution, development, progress are not only on her programme, these are her programme. For all things are rising, all worlds, all planets, all stars and suns. An ascending energy is in the universe, and the whole moves on with one mighty idea and anticipation. The aspiration in the human mind and heart is but the evolutionary tendency of the universe becoming conscious. Darwin's great discovery, or the discovery which he brought into prominence, is the same as Galileo's—that the world moves. The Italian prophet said it moves from west to east; the English philosopher said it moves from low to high. And this is the last and most splendid contribution of science to the faith of the world.

The discovery of a second motion in the earth has come into the world of thought only in time to save it from despair. As in the days of Galileo, there are many even now who do not see that the world moves —men to whom the earth is but an endless plain, a prison fixed in a purposeless universe where untried prisoners await their unknown fate. It is not the monotony of life which destroys men, but its pointlessness; they can bear its weight, its meaninglessness crushes them. But the same great revolution that the discovery of the axial rotation of the earth effected in the realm of physics, the announcement of the doctrine of Evolution makes in the moral world. Already, even in these days of its dawn, a sudden and marvellous light has fallen upon earth and heaven. Evolution is less a doctrine than a light; it is a light revealing in the chaos of the past a perfect and growing order, giving meaning even to the confusions of the present, discovering through all the deviousness around us the paths of progress, and flashing its rays already upon a coming goal. Men begin to see an undeviating ethical purpose in this material world, a tide, that from eternity has never turned, making for perfectness. In that vast progression of Nature, that vision of all things from the first of time moving from low to high, from incompleteness to completeness, from imperfection to perfection, the moral nature recognizes in all its height and depth the eternal claim upon itself. Wholeness, perfection, love—these have always been required of Man. But never before on the natural plane have they been proclaimed by voices so commanding, or enforced by sanctions so great and rational.

Is Nature henceforth to become the ethical teacher of the world? Shall its aims become the guide, its spirit the inspiration of Man's life? Is there no ground here where all the faiths and all the creeds may meet—nay, no ground for a final faith and a final creed? If all men could see the inner meaning and

aspiration of the natural order should we not find at
last a universal religion—a religion congruous with
the whole past of Man, at one with Nature, and with
a working creed which Science could accept ?

The answer is a simple one: We have it already.
There exists a religion which has anticipated all these
requirements—a religion which has been before the
world these eighteen hundred years, whose congruity
with Nature and with Man stands the tests at every
point. Up to this time no word has been spoken to
reconcile Christianity with Evolution, or Evolution
with Christianity. And why? Because the two are
one. What is Evolution? A method of creation.
What is its object? To make more perfect living
beings. What is Christianity? A method of creation. What is its object? To make more perfect
living beings. Through what does Evolution work?
Through Love. Through what does Christianity
work? Through Love. Evolution and Christianity
have the same Author, the same end, the same spirit.
There is no rivalry between these processes. Christianity struck into the Evolutionary process with no
noise or shock; it upset nothing of all that had
been done; it took all the natural foundations precisely as it found them ; it adopted Man's body, mind,
and soul at the exact level where Organic Evolution
was at work upon them; it carried on the building by
slow and gradual modifications; and, through processes governed by rational laws, it put the finishing
touches to the Ascent of Man.

No man can run up the natural lines of Evolution
without coming to Christianity at the top. One holds
no brief to buttress Christianity in this way. But

science has to deal with facts and with all facts, and the facts and processes which have received the name of Christian are the continuations of the scientific order, as much the successors of these facts and the continuations of these processes—due allowances being made for the differences in the planes, and for the new factors which appear with each new plane—as the facts and processes of biology are of those of the mineral world. We land here, not from choice, but from necessity. Christianity—it is not said any particular form of Christianity—but Christianity, is the Further Evolution.

"The glory of Christianity," urged Jowett, "is not to be as unlike other religions as possible, but to be their perfection and fulfilment." The divinity of Christianity, it might be added, is not to be as unlike Nature as possible, but to be its coronation; the fulfilment of its promise; the rallying point of its forces; the beginning not of a new end, but of an infinite acceleration of the processes by which the end, eternal from the beginning, was henceforth to be realized. A religion which is Love and a Nature which is Love can never but be one. The infinite exaltation in quality is what the progressive revelation from the beginning has taught us to expect. Christianity, truly, has its own phenomena, its special processes, its factors altogether unique. But these do not excommunicate it from God's order. They are in line with all that has gone before, the latest disclosure of Environment. Most strange to us and new, most miraculous and supernatural when looked at from beneath, they are the normal phenomena of altitude, the revelation natural to the highest height. While Evolution never deviates from its course, it assumes new developments at every stage of the Ascent; and here, as the last and highest, these specializations, accelerations, modifications, are most revolutionary of all. For the evolving products are now no longer the prey and tool of the Struggle for Life—the normal dynamic of the world's youth. For them its appeal is vain; its force is spent; a quicker road to progress has been found. No longer driven from below by the Animal Struggle, they are drawn upward from above; no longer compelled by hate or hunger, by rivalry or fear, they feel impelled by Love; they realize the dignity reserved for Man alone in evolving through Ideals. This development through Ideals, the Perfect Ideal through which all others come, are the unique phenomena of the closing act—unique not because they are out of relation to what has gone before, but because the phenomena of the summit are different from the phenomena of the plain. Apart from these, and not absolutely apart from these—for nothing in the world can be absolutely apart from anything else, there is nothing in Christianity which is not in germ in Natture. It is not an excrescence on Nature but its efflorescence. It is not a side track where a few enthusiasts live impracticable lives on impossible ideals. It is the main stream of history and of science, and the only current set from eternity for the progress of the world and the perfecting of a human race.

We began these chapters with the understanding that Evolution is history, the scientific history of the world. Christianity is history, a history of some of the later steps in the Evolution of the world. The **continuity** between them is a continuity of spirit;

their forms are different, their forces confluent. Christianity did not begin at the Christian era, it is as old as Nature; did not drop like a bolt from Eternity, came in the fulness of Time. The attempt to prove an *alibi* for Christianity, to show that it was in the skies till the Christian era opened, is as fatal to its acceptance by Science as it is useless for defence to Theology. What emerges from Nature as the final result of Creation is the lower potentiality of the same principle which is the instrument and end of the new Creation.

The attempt of Science, on the other hand, to hold itself aloof from the later phases of developments which in their earlier stages it so devotes itself to trace, is either ignorance or affectation. For that Altruism which we found struggling to express itself throughout the whole course of Nature, what is it? " Altruism is the new and very affected name for the old familiar things which we used to call Charity, Philanthropy, and Love." [1] Only by shutting its eyes can Science evade the discovery of the roots of Christianity in every province that it enters; and when it does discover them, only by disguising words can it succeed in disowning the relationship. There is nothing unscientific in accepting that relationship; there is much that is unscientific in dishonoring it. The Will behind Evolution is not dead; the heart of Nature is not stilled. Love not only was; it is; it moves; it spreads. To ignore the later and most striking phases is to fail to see what the earlier process really was, and to leave the ancient task of Evolution historically incomplete. That Christian

[1] Duke of Argyll, *Edinburgh Review*, April, 1894.

development, social, moral, spiritual, which is going on around us, is as real an evolutionary movement as any that preceded it, and at least as capable of scientific expression. A system founded on Self-Sacrifice, whose fittest symbol is the Leaven, whose organic development has its natural analogy in the growth of a Mustard Tree, is not a foreign thing to the Evolutionist ; and that prophet of the Kingdom of God was no less the spokesman of Nature when he proclaimed that the end of Man is " that which we had from the beginning, that we *love.* "

In the profoundest sense, this is scientific doctrine. The Ascent of Man and of Society is bound up henceforth with the conflict, the intensification, and the diffusion of the Struggle for the Life of Others. This is the Further Evolution, the page of history that lies before us, the closing act of the drama of Man. The Struggle may be short or long ; but by all scientific analogy the result is sure. All the other Kingdoms of Nature culminated ; Evolution always attains ; always rounds off its work. It spent an eternity over the earth, but finished it. It struggled for millenniums to bring the Vegetable Kingdom up to the Flowering Plants, and succeeded. In the Animal Kingdom it never paused until the possibilities of organization were exhausted in the Mammalia. Kindled by this past, Man may surely say, " I shall arrive." The Further Evolution must go on, the Higher Kingdom come—first the blade, where we are to-day ; then the ear, where we shall be to-morrow ; then the full corn in the ear, which awaits our children's children, and which we live to hasten.

FINIS.

COSIMO is a specialty publisher of books and publications that inspire, inform and engage readers. Our mission is to offer unique books to niche audiences around the world.

COSIMO CLASSICS offers a collection of distinctive titles by the great authors and thinkers throughout the ages. At **COSIMO CLASSICS** timeless classics find a new life as affordable books, covering a variety of subjects including: *Biographies, Business, History, Mythology, Personal Development, Philosophy, Religion and Spirituality,* and much more!

COSIMO-on-DEMAND publishes books and publications for innovative authors, non-profit organizations and businesses. **COSIMO-on-DEMAND** specializes in bringing books back into print, publishing new books quickly and effectively, and making these publications available to readers around the world.

COSIMO REPORTS publishes public reports that affect your world: from global trends to the economy, and from health to geo-politics.

FOR MORE INFORMATION CONTACT US AT
INFO@COSIMOBOOKS.COM

❋ If you are a book-lover interested in our current catalog of books.

❋ If you are an author who wants to get published

❋ If you represent an organization or business seeking to reach your members, donors or customers with your own books and publications

COSIMO BOOKS ARE ALWAYS
AVAILABLE AT ONLINE BOOKSTORES

VISIT COSIMOBOOKS.COM
BE INSPIRED, BE INFORMED

CPSIA information can be obtained at www.ICGtesting.com
Printed in the USA
BVOW08s1151210515

401111BV00001B/87/P